# A SHORT HISTORY OF FARMING

IN

# NEW ZEALAND

GORDON McLAUCHLAN

# A Short History of Farming in New Zealand

Published in 2020 by David Bateman Ltd
Unit 2/5 Workspace Drive, Hobsonville, Auckland 0618, New Zealand
www.batemanbooks.co.nz

ISBN 978-1-98-853801-3



Cover images: Gumboots GH017433 and Ko (digging implement) ME001745 images supplied courtesy of Te Papa. All other images via iStock.
Book design: Alice Bell
Printed in China by Everbest Printing Co. Ltd

# Contents

# Introduction

For eight hundred years — from the first discovery and habitation of Aotearoa/New Zealand by Polynesians — until the 1970s, the tangata whenua and later European immigrants were physically and emotionally dependent on and sustained by the good earth, living against rural landscapes or on the coast where they worked directly for their food, clothing and housing.

When the first immigrants arrived in the twelfth century from what had been their tropical homeland, they found their new country had abundant flora but narrow in range and, of the tropical food plants they brought with them, only the kūmara thrived and only in some parts of the country after careful nurturing. It became their most favoured source of carbohydrate. Fauna was also abundant but the Polynesian pig did not accompany these first arrivals to provide them with an animal to farm. So as well as farmers, they became hunters and gatherers among the rich sources of bird and marine life.

It is a common mistake to believe that people in such societies lived — as seventeenth century philosopher Thomas Hobbes claimed — 'solitary, poor, nasty, brutish and short' lives. Welsh prehistorian, Alasdair Whittle, among others, has emphasised that early European hunters and gatherers were, in fact, technically proficient in their management of resources and lived intensely social lives. And this would certainly apply to Māori.

In the last quarter of the nineteenth century, New Zealand became a British offshore farm. Many of the trees that once luxuriously clothed the land were cut down and replaced by grass for pastoral farming and by northern temperate zone food plants that thrived here at least as well as where they came from. What wasn't eaten or processed here was shipped away in 'home boats'.

Māori lived in whānau groups attached to hapū and within iwi, and emotionally many still do. European migrants at first lived in rural isolation before many gathered in towns and small cities to service the people working the good earth. And in their backyards, they worked the soil themselves. An academic visiting Victoria University of Wellington in the 1960s said gardening was an important part of New Zealand culture at that time. Even after cities grew quickly following the Second World War, New Zealanders remained culturally a rural people.

During the two decades that followed the war, the major book publishing successes were almost entirely tales that reflected life in the countryside. I have a library of books, many of them humorous, that were the best-sellers of the 1950s and 1960s — very few were not about country life. A resurgence in the 1950s and 1960s of new fiction from writers such as Janet Frame, Maurice Gee and Ian Cross was still mainly about the country or small towns. The country's biggest newspaper in the country's most densely packed urban region, *the New Zealand Herald*, did not have a business editor until late in the 1960s, and even then the agricultural editor remained a major force in the editorial department.

Since the 1980s, the national change to an urban culture has been dramatic. Not long ago, newspapers carried a New Zealand Press Association story about a farmer who was 'rammed' against the 'rails of a pen' by a bull immediately after he had inserted an ear tag. According to a rescue helicopter spokesman, the farmer 'ended up with one of the horns between his legs which missed everything' — 'everything' being a euphemism I'm sure the farmer enjoyed once he recovered from 'the serious bruising to his lower back and spine'. But what made me laugh out loud was the last sentence of the story: 'It was unclear why the bull rammed the man.'

A bull is one kind of ruminant, the sort that chews its cud, but it is not another, reflective, rational kind of being the *Concise Oxford Dictionary* defines as 'given to meditation'; so I'm sure the cause of his attack was also unclear to the bull behind a red mist of fear and anger.

The journalist who wrote the story was clearly a city-dweller and if you live in a city nowadays, especially a biggish one like Auckland, you see the grass in such small patches and forget, if you ever knew, that it is the stuff cud and then milk and meat is made of, the stuff that largely supported the economy of this country for about 150 years.

No one of my generation in any walk of life could possibly have escaped the knowledge at that time that New Zealand earned its living in the world by selling wool, meat, dairy produce, and apples and pears. You didn't just learn it at school; it pressed on your life.

The seas of grass lapped into the suburbs, straggling around sagging wire fences. We played football after school among the sheep's 'tollies' and the cowpats. If cuts on arms and legs sometimes swelled with pus, stinging hot poultices drew the poison out. I feel a great unease now as some of the finest grassland in the world recedes, yielding to urban subdivisions and their barren, monotonous houses that may one day spawn barren, monotonous children.

From eleven through fourteen I spent summer holidays and many weekends milking cows and hay-making in shorts and sandals and not much else: red skin peeled and, after a week or two of each summer, left a heavy brown shell as well as unseen scars that blossom now as basal cells.

On the twice-daily train between the Hutt Valley and Wellington College, the stink would punch into my nostrils as I went past the Ngāūranga freezing works at the foot of the gorge, but I didn't recoil then as I would now in a more sanitised world. A few years later, for three summer seasons, I worked, enveloped in that pong but not noticing it any more, on the end of a butchers' chain, fit but too small for the chain itself, on which powerful men sweated through eight-hour days at what they called 'piece rates'. Paid by the carcase throughput, the contract chains went only as fast as the slowest links, so the weak and the slack were soon prised out.

For overtime at night, we worked in teams of four folding lambs' legs through the brisket and pulling the shortened carcase into a muslin bag before it was frozen. Telescoping, it was called. Designed during the war to save shipping space, it survived the war for a decade or more. They were stumpy lambs then, swathed in layers of white fat, not like the lean carcases of today. We students worked with common men who often turned out to be not so common at all. Many were autodidacts who had been reading in their ships' bunks for years before they deserted to live here.

Full-time university students were a rarity: some of the sons and daughters of the privileged, unenvied, who missed out on the richest part of our education. Many students not in the freezing works over summer mostly worked on the wharves or in the woolstores. On the hottest days at Ngāūranga, we would sometimes swim off the rocks during the tea break, checking first which way the sea was sweeping the blood and guts from the effluent pipe, so we could swim up-tide and avoid the unpleasant viscera.

All this was within the city. City workers who idled away a sunny lunchtime with a walk on the wharves would see almost nothing but wool bales, carcases and butter boxes being craned aboard the 'home boats'.

That's all gone now, and it went very quickly from the 1970s. Town and country had not always lived easily together, but their lives had always interlocked.

For several years in the 1960s, I spent a lot of time driving around New Zealand on assignment for *The New Zealand Weekly News*, most often called 'The Auckland Weekly', and for the *New Zealand Journal of Agriculture* of which I was editor for four years. I gloried in the New Zealand countryside: from the backblocks of Canterbury and the rough, broken landscapes of Molesworth Station and Central Otago (worked by the hardest of men) to the kempt, almost manicured orchards and stud farms in places like the Wairarapa, Manawatū and Bay of Plenty.

I also enjoyed farmers: often taciturn, easy-mannered men with an earthy sense of humour, often slow to speak but firm to the point of obduracy in their opinions once expressed; always sizing up townies, looking for the mental and moral flab they were pretty sure they would find sooner or later.

I remember the competence of their wives at what is now called multi-tasking but had no name then. They just did all those things because there was no one else to do them and few machines to give them respite in kitchen and laundry. And the children, many of whom fled the farm for the professions in town because it was such an unremittingly demanding life, but so often wearing the confidence and maturity the hard physical work had lent them. I smile when I think of the gorgeous young woman home on holiday from university to the family Romney stud, emphasising to a visiting group of foreign farmers that the ram that had just serviced a ewe with such alacrity in the shed was typical. 'All Dad's rams jump like that,' she said proudly.

But I understood this taciturnity often enough glazed over ignorance based on the assumption that merely living in the country provided a base for wisdom, that the absence of tall buildings and crowds of people was in itself a spiritual blessing. I once drove from Iowa to a neighbouring state, hemmed in on both sides of the road for mile after mile by a wall of maturing corn plants and realised that factory farming was indeed as enslaving as an assembly line. Huge corporate dairy farms are as far from bucolic bliss as the cornfields of Iowa. When farming becomes just an industry it seems to bewilder many farmers. Some psychologists are rediscovering that life in the country can be therapeutic, but only if you are close to the good earth and its trees and hedges and especially if you are personally working with it to recreate the cycle of life.

As I drove around the country in the 1960s and 1970s, the history of New Zealand economic life hung in the air and was stamped on the land by dilapidated cowsheds, shearing sheds, barns, and old milk-stands at the gate, replaced by old reapers, traction engines, tractors, small trucks, partly cannibalised and waiting still for a decent grave in case their bones could still be picked. Farmhouse kitchens and laundries were being remodelled at last and wives were relishing the appliances if not quite trusting them yet to properly clean and dry, so the coppers and lead-lined tubs were still in their place in the old washhouse, 'in the meantime', which meant for years.

In 1986, the government abruptly pulled all the financial subsidies that had supported research and development (and had been abused by a previous government). The belief was that farming was not going to be the force in the future it had been in the past.

A visiting English agricultural journalist, Claire Powell, wrote to her home paper: 'Farmers are reluctantly reading the writing that has been on the wall for some time — that their traditional farm produce is in a state of worldwide oversupply; their home market of 3.1 million is minute and their God-given rights to vital overseas markets are weakening.' She was partly right — as we know now more than thirty years and nearly two million more people later — but only partly, as growing populations have put food producing back on the menu.

She also wrote with the portentousness we all felt: 'The main victims of the present crisis are the enthusiastic, heavily mortgaged young men, often on as much as eighty-five per cent borrowed money. They are being the first to be forced from the land. There are few willing or able to fill their shoes under present circumstances. Only time will tell if a vacuum of competent farmers is being created.'

Time has told. A measure of the recent achievement of farming and the confidence its success had built in its future was the increase in the value of rural land: it doubled between 1990 and 2000, and doubled again between 2000 and 2005.

A cultural change towards more intensive urbanisation began, but only gradually, after the Second World War. Our literature is rich with stories of remoteness and isolation, living hard lives in the bush and distant paddocks. A shearer became an international celebrity for devising a more economic technique.

What is now pressing on the country is a need to contain and reduce greenhouse gas emissions from pastoral farming and a need to continue reassessment of land use. As world tastes and attitudes towards food change, land needs to be more productive than it has become by simply running animals over grassland. Horticulture is expanding and will continue to do so in the decades ahead. But, as the nature of farming changes in fundamental ways, it is important to remember that the history of this country — since the beginning of human habitation — lies deeply rooted in the good earth which has conditioned what we have been and what we have become, culturally and economically.

The history of this
country — since the
beginning of human
habitation — lies
deeply rooted in the
good earth which has
conditioned what we
have been and what we
have become.

## Chapter One

# The First Farmers

Polynesians — fishers and gardeners — were the nomads of the tropical Pacific and for hundreds of years had much of it to themselves, on the move in their wonderful little ships, pausing here and there, maybe for decades, maybe for centuries. The reasons for their journeys of exploration were complex but must have included some deep cultural longing to move on across the sea, prompted by the smallness of their island homes constricting population growth and exacerbating political and religious tensions. And the sea, after all, was as much their home as the land.

They became experienced and adept at the business of packing and leaving, and then arriving somewhere new, always a major economic project in the islands of departure. They took with them their pigs and their plants all the way across the Pacific from the west to the east — until they came to temperate zone New Zealand, mysteriously without pigs. Of the permanently settled Polynesian islands, only Rapa Nui (Easter Island) and New Zealand lie below the Tropic of Capricorn with New Zealand even farther to the south. On some tropical atolls, the soil would not support some of the richer food plants, breadfruit, for example, although the ubiquitous coconut seldom, if ever, failed them.

So the first Polynesian settlers came to New Zealand from the Arcadian islands of the tropical Pacific — the Society Islands, almost certainly via the Cook Islands — into an alien environment, carrying food plants that failed to survive at worst and barely clung to life at best in the cooler climate. They were quickly and comprehensively dispossessed a few centuries later by Europeans who found here a land and climate ideally suited to the crops and animals they brought from their own, more distant, cool Hawaiki.

The early migrants to New Zealand must have been dismayed to find the colder temperatures would not allow them to enjoy the prodigality of the coconut, the banana or the breadfruit. In the high, warm islands of Hawaii, these tropical fruits and pigs supported a population of at least 300,000, about three times that of Aotearoa. Of the other plants they brought here, the kūmara would flourish only in northern coastal regions but grow most years as far south as Banks Peninsula if carefully tended. The seed tubers needed to be stored through the cold months for replanting in the spring.

Taro grew only in some parts of the north, needed much attention and even then took a long time to mature.

The bottle gourd (hue) was propagated from seed, the only Māori staple that was, and it was eaten as a young plant in the period before the kūmara was ready. Many of them were also allowed to mature and then hardened in the sun or by fires and used as vessels to hold water, oils and other liquids, as well as birds and fish preserved in fat. Another use was as floats. The gourd's flesh was thin, like the European marrow, and of less nutritional value than the kūmara, taro or yam.

The yam (uhi or uwhi) was grown extensively in the Bay of Plenty and was common on the East Coast and other warm coastal regions of the north where it needed much attention; but it did not thrive elsewhere. It languished and then disappeared after the introduction of the potato.

The tropical cabbage tree (tī pore), also believed to be brought in by early Polynesian immigrants, was planted by Māori near settlements but never became prolific. The most prized of a number of cabbage tree varieties growing here, tī pore required a lot of labour to dig out, cut and cook its large elongated root, and it was eaten as a delicacy rather than a staple food. Local plants such as karaka trees were planted by Māori near their settlements for their fruit but not cultivated in plantations, just as flax (harakeke, *Phormium tenax*) was planted within easy range of whānau.

The paper mulberry (aute) did not grow well and evidence suggests that attempts were first made to substitute the felted clothing material from the aute with the bark of the lacebark (*Hoheria* species) before the wonders of flax were discovered. Māori women became expert at refining harakeke and weaving it into fine fabrics as well as ropes, nets, baskets and mats.

The livestock most migrating expeditions would have set out with included pigs, fowls, dogs (kurī) and rats (kiore). Pigs and fowl never made it all the way — or at least fowl may have but were less valued as food by the newcomers than the meaty and easily hunted land and sea birds that flocked in huge numbers throughout the country. Dogs were the only animal they domesticated.

Species of moa, kiwi and other plump land birds unused to predators were easy pickings and became rare within a relatively short time, not only because of overhunting but also because of environmental changes such as deforestation.

Thus, the earliest arrivals from Polynesia became hunters and gatherers rather than the gardeners they had been on their islands of origin. Species of moa, kiwi and other plump land birds unused to predators were easy pickings and became rare within a relatively short time, not only because of overhunting but also because of environmental changes such as deforestation wrought by these first humans to arrive. Seals and a variety of fish, shellfish and seabirds also became everyday foods. The basic carbohydrate came from aruhe, the rhizome of the bracken fern, *Pteridium esculentum*, which was available in most parts of the country. It was dug up and ground into a rough powder, and easily stored.

Although the forest was a source of timber for canoes, buildings, weapons and some food — edible berries, young palm fronds, the bulbous upper stem of the nīkau palm, roots and birds — it could not and did not sustain dense populations. It was a narrow range of flora compared with that of the temperate countries of the northern hemisphere. Therefore, destruction by burning was not lamented by Māori intent on encouraging new fern growth and the development of kūmara gardens as the fertility of old ones was expended.

Because of the difficulty of controlling burn-offs, many fires would have spread far beyond the intentions of Māori gardeners. One estimate is that about half the area covered by forest when the Polynesians arrived had been destroyed by the time Pākehā came to settle. Some forests, especially in low-rainfall regions on the east coast, never regenerated beyond native grasses and scrub. It was on this land that the first Pākehā farmers ran their sheep.

Burning and tree-felling with iron tools continued after the prolific potato was introduced by Europeans who, intent on pastoral farming, became even more efficient at denuding the land of forest than their predecessors. Before the end of the nineteenth century reforestation had become a political issue.

The profligacy of Māori with resources is not hard to understand. Land birds and bracken were readily available in a country that seemed huge, and was, in fact, bigger than all the other islands of Polynesia put together. Moa, other birds, and seals were easy to catch and kill and for a long time must have seemed an infinite resource until they became, in some cases, extinct or close to extinction.

There were nine or ten species of moa in both islands and most offshore islands. The largest of these ratite birds was the *Dinornis novaezealandiae* in the North Island and *Dinornis robustus* in the South Island, both more than two metres tall and heavily built. Others were smaller, down to the size of a turkey, and all were hunted by Māori as a rich source of food. They had been on the ground for a long time, the only ratites with no trace of any former wing formation, and their only predator before humans arrived was the huge native eagle.

They were easy game these grounded birds. Even kiwi made the hāngi. In 1890, the Anglo-Indian writer, Rudyard Kipling, visited New Zealand and wrote of his visit here: 'From Wellington I went north towards Auckland… At one of our halts I was given for dinner a roast bird with a skin like pork crackling,

'I was given for dinner a roast bird with a skin like pork crackling, but it had no wings nor trace of any. It was a kiwi — an apteryx. I ought to have saved its skeleton, for few men have eaten apteryx.'

but it had no wings nor trace of any. It was a kiwi — an apteryx. I ought to have saved its skeleton, for few men have eaten apteryx.'

As the land birds dwindled from hunting and an environment changed by humans, gardening increased in importance for Māori. While they had access to native pūhā (sow thistle) and other vegetables and fungi, the food most commonly cultivated and most prized was undoubtedly the tasty and nutritious kūmara. Around the world, all populations had their own staple energy foods available from grains such as rice, wheat, oats, millet and maize, and from root vegetables such as potatoes, yams, sweet potatoes, cassava and taro.

By the time James Cook arrived, the shortage of land birds meant Māori — estimated by a member of his crew as about 100,000 — were farmers again with the main centres of population gathered around the cultivation of the kūmara in the coastal regions of the North Island, or inland around the thermal areas of the Bay of Plenty and along the fertile banks of the Waikato and Whanganui rivers. Some land in the north of the South Island was also farmed.

Much labour and ritual were invested in the annual kūmara crop. In early spring, members of each hapū attacked the area to be planted with the kō, or digging stick. Most Polynesian populations had a version of the digging stick, also called kō, or something very close to that. It was used throughout New Zealand, mostly made of maire or mataī hardened by fire or by ageing, varying in size and often with modifications of design and some were decorated with carvings at the handle end. It was around two metres long with a sharpened blade at the bottom eight to ten centimetres wide.

Most kō had a footrest lashed into a slot on the side of the blade, enabling diggers to use their foot to help plunge the blade deeply into the ground. Iron spades were in high demand for digging post-holes, for trenching and other earthworks after they were introduced by Europeans, but the kō lasted for many years among Māori gardeners.

Following the diggers would come the women and children, down on their haunches chopping up the turned sods with smaller sticks, some shaped like paddles, most commonly called kāheru. They would construct quincuncial heaps of friable soil ready for the seed tubers. If the soil was not naturally light, friable or porous enough for the kūmara then sand, shells or fine shingle from a riverbank or a shingle pit would be mixed in. And light shingle with vegetation might be used as a mulch for protection of the growing tendrils against wind or frost.

Edward Shortland, who held administrative posts in New Zealand for nearly fifty years from 1841, wrote that in parts of the Waikato near old kūmara gardens 'the traveller frequently meets with large excavations, from twenty to thirty feet in depth, the gravel pits one is accustomed to see in England near public roads'. Heavy shingle and large stones were used to build small walls, and woven brush fences were erected for protection against wind and foragers such as the pūkeko, kiore and, later, pigs.

Like the populations of other farming cultures, Māori ritualised the cultivation of their basic energy food, the kūmara, from the selection of the site of a garden, through preparation of the soil and planting in spring to a harvest-festival style of digging the mature tubers in the late autumn. Each step was accompanied by appropriate gestures and incantations from tohunga (priests, experts). The phases of the moon were taken into account, the stars observed, soil temperature tested and, among some hapū, a stone figure of the god Rongo anointed where it stood in the field. Without care the tubers may be planted at the wrong time for early growth or harvested at the wrong time for storage.

Early missionary William Colenso, who spent most of his life in New Zealand from 1833, wrote that the 'extensive tribal kumara plantations' were worked by tribal members 'in a compact body, chief and slave, keeping time with their songs, which they also sang in chorus'. After the intensive work of preparing and planting, the gardens would be tended in a more desultory way, but according to early Pākehā observers they were meticulously weeded and the hillocks hoed up throughout the late spring and the summer.

Before the main crop was dug late in March or in April, some of the larger immature tubers were fingered free from the underground stems. These were skinned, dried on mats, turned over in the sun and eaten as a late summer treat or stored. Sites such as Anaura Bay on the East Coast were ideal for farming with the tribe able to avoid seasonal moves for other food. The soil and the climate were right and the sea and coastline fruitful.

Soon after her arrival, Captain James Cook's *Endeavour* sailed north along the east coast, and anchored in a bay where the ship's surgeon, William Brougham Monkhouse, wrote in his journal:

> *We had an opportunity to examine their Cultivations more at leasure today and found them very far to surpass any idea we had formed of them. The ground is compleatly cleared of all weeds ... with as much care as that of our best gardens. The Sweet potatoes [kūmara] are set in distinct little molehills which ranged in some in straight lines, in others in quincunx. In one Plott I observed these hillocks, at their base, surrounded with dried grass.*
>
> *The Arum [taro] is planted in little circular concaves, exactly in the manner our gardeners plant Melons. ... The yams are planted in like manner with the sweet potatoes: these cultivations are enclosed with a perfectly close pailing of reeds about twenty inches high. The natives are now at work compleating these fences. We saw a snare or two set upon the ground for some small animal [probably kiore, the Pacific rat] ...*
>
> *It is agreed that there are a hundred acres of ground cultivated in this Bay — the soil is light and sandy in some parts — on the sides of the hills it is a black good mold. We saw some of their houses ornamented with gourd plants in flower. These with the Yams, sweet Potatoes and Arum are so far as we yet know, the whole of what they Cultivate.*

The tohunga was the repository of knowledge, the man who knew the garden would last three seasons and then, in its exhaustion, be given over to fernroot, the plentiful staple food that would grow there before it went back to mānuka and could be burnt for a new kūmara patch a few years later.

The correct incantations and karakia (a prayer usually chanted) gave the Māori gardeners a sense of control over their destiny, invoking the spirits of their ancestors to intercede with the gods on their behalf. These invocations to the gods of food growing, Rongo and Pani, were deeply poetic. William Colenso studied them and found translation extremely difficult because of allusiveness, ellipsis, metaphor and archaisms.

Māori wove an elaborate and elegant mythology around the culture of the kūmara, and the poems they spoke varied in style and content among iwi. They and their more recent Polynesian ancestors had been isolated from any other ethnic group for at least a thousand years by the time they arrived in Aotearoa and had developed their own religion, complete with their 'Genesis'. Among New Zealand Māori, Rongo-marae-roa (Rongo of the Great Domain), or Rongo-nui (Great Rongo) was the god of peace and cultivated food, and sometimes refined down to the god of the kūmara. Rongo was the supernatural offspring of Rangi-nui the Sky Father and Papa-tū-ā-nuku the Earth Mother.

Peace and agriculture were wrapped together in Māori society. Wars were seldom fought during planting and harvesting. Even during the civil war of the 1860s, Māori warriors supporting other hapū would go home to attend to their seasonal agricultural duties.

Rongo and Tāne (god of the forest) and Tangaroa (god of the sea) are the most widely known among the deities of Polynesia. According to another Māori myth, a relative of Rongo called Pani gave birth to the kūmara and different varieties of the plant that evolved here were named as her children. According to ethnologist and writer Elsdon Best, Pani also takes the form of a male in some other iwi myths, and in some she is married to Tiki, the male fertility figure. The correct incantations and karakia invoked the spirits of their ancestors to intercede with the gods on their behalf.

The origins of the Polynesian kūmara are obscure and somewhat mysterious. Most botanists agree it came from the Americas and — with the fact of the prevailing easterly wind across the tropical Pacific — this was a reason some anthropologists supported the now largely debunked claim that Polynesians themselves originated in South or Central America. Such a claim was made by Norwegian anthropologist Thor Heyerdahl in his readable, 1948 best-seller, *Kon Tiki*.

The possibility is that some Polynesians, after centuries of sailing eastwards into the sun, reached South America and found an intimidating land mass very different from the smaller, sea-girt islands of their homelands. Also, people already lived there and were easily able to repel prospective settlers arriving in canoes carrying, say, thirty people. Some may have lingered long enough to carry the sweet potato and some other plants back to the Society Islands or the Marquesas.

Or there may be a germ of truth in the hypothesis that South or Central Americans did reach the Marquesas or some other island of Eastern Polynesia in their small boats and had the kūmara aboard.

Taro and gourds were planted about the same time as kūmara in the relatively few places where they could survive, the taro in rich, moist soil and carefully sheltered by brush fences against wind and cold. The leaves and stem, carefully cooked, were eaten as well as the root.

Techniques for storing kūmara were developed over the years, usually in deep subterranean pits, and thus gardens became larger and larger to produce tubers that could be safely stored to support hapū from one autumn to the next, and for spring planting. A number of varieties developed in New Zealand in various soils and microclimates. The most common had a tuber described by early arrival Joel Samuel Polack as 'being the size and shape of the finger and extremely farinaceous and nutritious'. The large red varieties were brought in later by Europeans, in one recorded case by an American whaler in 1819.

Māori living in the northern coastal districts of the North Island had access to abundant food resources, provided by their own farming, fowling and fishing skills, even after the large land birds had been plundered to extinction.

A wider range of economic activity was practised by hapū in their villages than was once thought. For example, archaeologists investigated and partly restored a village they called Kohika, which was built on an island surrounded by a lake in the midst of what was once a large swamp on the Rangitāiki Plains in the Bay of Plenty. The village was a substantial settlement on the edge of the water, not unlike a modern marina, and thriving before its cataclysmic end not long before the arrival of the first Europeans. Its inhabitants gardened, gathered wild vegetables and took advantage of abundant fish and bird resources.

The village was also was a pivotal trading centre for wood and stone (mainly obsidian) because of its easy access to a network of inland waterways, to the sea and to higher land. It was hierarchical with a chiefly house in pride of place, and rich enough for some division of labour to allow for the specialisation of artisans. Its site was economically advantageous but also perilous and it was abandoned after a flood cut it off from its access to the lake, rivers and the sea.

The northerners learned to live with seasonal rhythm, planting, harvesting and storing, trapping birds and preserving any surplus in the birds' own fat, catching eels, combing the foreshore for its rich resources, netting and hooking pelagic fish and drying them for storage. Green vegetables and the fruits of forest trees were readily available. As Joseph Banks reported on the *Endeavour*'s visit: 'Tillage, weaving and the rest of the arts of peace are best known and most practised in the north eastern parts…'

Ethnologist David Simmons wrote of the spring and summer life of Māori at the time of Cook's first visit in 1769:

> *At Anaura Bay, Wharataewa and the Bay of Islands , the people were living in their base agriculture area either because of imminent attack … or because their access to sea resources was close and fruitful. In some cases then, the move to seasonal stations need not have yet taken place. This period in the agricultural cycle involved planting the crop, then living mainly on fish and fern root which often also involved some form of dispersal to coastal areas or along coastal areas.*

He said that in Wharataewa pā the people had a flax seine net 73.16 metres long, requiring a number of people to pull it. When conditions were suitable along the sandy beach the return on such a net would enable the people to stay in larger groups. Seine nets more than a kilometre long were reported along the sandy coastal stretches of the East Coast bays, Tokomaru, Anaura and Tolaga.

The abundance gave northern Māori the advantages of economic specialisation and more leisure to practise arts and crafts, and because the environment was relatively congenial, the highest concentrations of population were in the coastal northern half of the North Island (as it is today), and by the time Europeans arrived this was imposing political pressures on land occupation, and had increased the tempo of tribal conflict.

Life in the north of the South Island and on the shores of inland lakes and large rivers in the North Island may also have been relatively comfortable during the warmer cycle of years believed to have existed during the early times of Māori occupation, especially in the fifteenth century. But it became harder in the far south and in inland regions of the North Island where a continental-type climate prevailed. When Europeans arrived the South Island, coastal regions teemed with fish, seals and bird life, but other foods were harder to find. A botanist with Cook on his second voyage, Georg Forster, thought Cook Strait a miserable place, cold and swept by southern and westerly winds, and populated by villagers who smelled of fish and rancid oil.

Cook left potatoes and other plants and he and subsequent visitors, including the French explorers and the whalers and sealers, continued to provide Māori of this southern temperate zone with gifts of plants and animals from the northern temperate zone. By the turn into the nineteenth century, the potato had replaced the kūmara as the staple energy food. It could be planted and harvested between the planting and harvesting of the kūmara as well as over a much wider region and, in some places, could provide two crops a year. Māori were nothing if not pragmatic. Although some conservative groups continued to practise agriculture in the old way, most were bedazzled by the new plants, tools and animals. And settlers found that, by and large, the livestock and plant stock they brought with them did better here than in Britain.

No one seems to know who first brought peaches to New Zealand but the stone-fruit, native to China, must have arrived in the early years of the nineteenth century, probably introduced by the first missionaries or possibly whalers. By the time the first European settlers arrived in the North Island 'Māori peaches' were thriving, especially along the rivers. Māori relished this fruit and planted the trees widely.

Māori had long made sophisticated use of harakeke, *Phormium tenax*, most commonly known later as New Zealand flax, for both clothing and cordage, and when the plant's value was recognised by Europeans it was adopted for domestic use and became a significant export in the nineteenth and early twentieth centuries. Products based on flax also became a significant export in the nineteenth and early twentieth centuries, but the industry never fulfilled its promise because it was extractive rather than based on cropping of selected varieties. The plant and its many varieties grew abundantly in swamps, so abundantly and so accessibly that Māori had not needed to plant it as a crop while using it widely for clothing and cordage.

When export demand first built up for the product in the nineteenth century, the processing expertise of Māori women was used. Their traditional ability enabled them to produce cloth for clothing so flexible and fine it was much admired by early European visitors.

As exports increased, mechanisation was needed to lift production. Export volumes and receipts reached 150 tons, worth £4674, as early as 1855, but then varied from year to year, reaching 22,653 tons worth £595,684 (three times as much as cheese earned) in 1903.

Once mechanised, flaxmillers, operating under freehold or licensed rights, set up their plants near swamps, harvested and milled the various strains of the plant, indifferent to the various qualities of each — in strength, length and colour. They operated much as a sawmiller did in stands of native forest. Little if any attention was given to choosing the superior varieties and then breeding and cultivating them as was done with food plant crops.

As competition developed from Manila hemp and Mexican sisal, an attempt was made in the 1930s, through the Council of Scientific and Industrial Research, to support hybridisation research and the development of high-quality strains. But it was too late. The Admiralty in Britain used New Zealand flax for some marine cordage and the industry scored a large contract to supply the United States Navy with rope early in the Second World War when production was averaging only 6000 tons a year.

By the 1950s, annual production was at 5000 tons as it continued its historic use in woolpacks for the New Zealand wool industry. Flax production faded soon afterwards in the face of the strength and flexibility of artificial fibres.

## Chapter Two

# Temperate Zone Crops and Animals

The weather was the main enemy of pre-European farmers, an enemy they fought with the rigorous canon of tapu, with elaborate litanies of hymn and incantation but also, mostly, with carefully acquired and memorised experience in coping with its caprices. Subsistence farming for a small population meant they were not seriously short of suitable land except in the coveted northern coastal regions in the late pre-European period.

But when Pākehā arrived with iron implements, potato, turnip, wheat, fruits and vegetables and pastoral and draught animals, and provided markets and a monetary transfer system, the weather became at once an ally and the infertility, the trees and steepness of so much of the land became the liability.

As Māori began something resembling commercial farming, their fields grew bigger and they became less scrupulous about rotating their use. They simply exhausted land and moved on. The competition for land and, to a lesser degree, competition for markets with produce, were factors in drawing the two competing races into war.

The first contact of Europeans with Māori brought sharp social change to Māori. They had previously been traders among themselves on a small scale — bartering kūmara from the north for greenstone from the south, kūmara and shellfish from the coast for birds and feathers from the inland forest. Obsidian and the work of artisans were also items of trade. But along the northern coasts of the North Island from the late eighteenth and early nineteenth centuries, Māori began to build their lives around the provisioning of ships, and supplying the small settlements of whalers, sealers and entrepreneurs from Sydney.

The first sealers had arrived at Dusky Sound only twenty-three years after Cook, in December 1792. Most Māori had long lived by the sea, but there is evidence that more of them moved down to the points of the coast the Pākehā regularly touched to swap pigs, potatoes and flax for clothes and metal artefacts. The race was still in its fullest vigour, sharply intelligent, rudely practical and industrious. It would continue to flourish, despite disease and the rapacious incursions of musket-deadly war parties, until the early 1860s.

The halcyon decades of the Māori farmer were the 1840s and 1850s. They still cultivated the kūmara in quantity, but the potato was cheaper and easier to grow and had virtually replaced fernroot. Whereas the potato could be lifted from the soil after a relatively short growing season, eaten immediately or stored for long periods, the kūmara had a longer growing season and fernroot needed tedious preparation before it could be eaten, and drying before it could be stored.

Their propensity for agriculture was given a fillip by the early missionaries who communicated that farming was the purest efficacy of the work ethic — a measurable return for earnest labour. The arch-missionary Samuel Marsden was from Yorkshire farming stock and believed the maxim 'as ye sow so shall ye reap' to be literally true. He considered farming as pleasing to God as to man. He came to New Zealand to re-establish contact after an act of utu (revenge) in 1809 for the flogging of a young Māori, Te Ara, by the captain of the *Boyd*. Te Ara's hapū boarded the vessel in Whangaroa Harbour and massacred the crew and passengers, which scared off Australian and British interest in New Zealand for several years.

Māori, with religious tensions surrounding agriculture, could easily understand what might be called the arable parables of the Bible — if not yet the pastoral ones. From his first visit in December 1814, the redoubtable Marsden encouraged the distribution of metal tools, so much more efficient than the stone and wooden implements of Māori. He brought with him the country's first horses, cattle and sheep accompanied by men with some knowledge of stockmanship.

For his first visit, Marsden set out from Sydney in the mission ship *Active*, on 28 November, with a party of thirty-five, including missionaries Thomas Kendall, William Hall and John King, eight Māori, two Tahitians — and a runaway convict who had stowed away and was discovered at sea. But it was the livestock that made the biggest impact on their arrival with 'one entire horse, two mares, one bull and two cows with a few sheep and poultry of different kinds'.

Among the passengers on the *Active* was John Liddiard Nicholas, a prolific writer. He wrote that when the animals were landed at Rangihoua, seven miles from the heads in the Bay of Islands, Māori were bewildered and amazed. 'Their excitement was soon turned into alarm and confusion, for one of the cows that was wild and unmanageable, being impatient of restraint, and unmanageable, rushed in among them and caused such violent terror through the whole assemblage that, imagining some preternatural monster had been let loose to destroy them, they all immediately betook themselves to flight.'

When Marsden mounted a horse and rode up and down the beach they followed him with staring eyes believing for the moment he was 'more than mortal'. (Thirty-four years later, horses and the pastoral animals had multiplied many times. A census in Auckland was taken and stock numbers in the district alone were: cattle, 137,204; sheep, 1,523,324; horses, 14,192; mules and asses, 122; goats, 11,797; pigs, 40,734.)

But if the missionaries were enthusiastic about encouraging Māori in agriculture, how able were the earliest among them to instruct them is debatable. In their Centennial Publication *The Farmer in New Zealand*, G.T. Alley and D.O.W. Hall wrote: 'The missionaries, although they vowed the intention of instructing the Maoris in the arts of civilization, were not themselves masters of the arts they taught — not at least of the art of farming.' But they were generally a tough and indefatigable cult of men, hewn from the granite of Anglo-Scottish evangelism.

A Ngāpuhi chief called Ruatara who had been round the world in European vessels had spent time on Marsden's farm at Parramatta in New South Wales. According to Marsden, Ruatara returned to the Bay of Islands, and 'laid out the grounds he intended to clear and cultivate, and marked out the ground for his men, having first enquired of me how much ground a man broke up in a day at Port Jackson'. He held out great hope for Ruatara as a progressive influence on Māori and their personal attachment seemed strong. The missionary was deeply saddened when Ruatara died suddenly from a mysterious illness soon after Marsden's return to Sydney from his first visit to the Bay of Islands in 1814. The powerful and able Hongi Hika, also a former visitor to Parramatta, took over the leadership and organisation of Māori agriculture in the north — but ultimately found warfare much more rewarding and his musket-toting warriors became the scourge of the North Island in the 1820s.

Māori quickly became proficient farmers of the new plants at their disposal. The first Pākehā visitors were fed by them. By 1830 they were involved in export trade. In that year twenty-eight ships (averaging 110 tons) made fifty-six voyages between Sydney and New Zealand. Māori-grown potatoes were the cargo on those ships, soon to be followed by Māori-grown and milled grain. Also that year, an agricultural mission school was set up in Waimate North, twelve miles inland from Paihia. Five years later it was visited by naturalist Charles Darwin, who wrote: 'On an adjoining slope fine crops of barley and wheat were standing in full ear; and in another part fields of potatoes and clover. … There were large gardens with every fruit and vegetable which England produces.'

The first plough was brought in by pioneer Anglican missionary John Butler and used at Kerikeri. A Marsden protégé, Butler wrote sanctimoniously in his diary:

> *On the morning of Wednesday the 3rd of May 1820, the agricultural plough was for the first time put into the land of New Zealand at Kiddie Kiddie [Kerikeri] and I felt much pleasure in holding it after a team of six bullocks brought down by the Dromedary. I trust that this auspicious day will be remembered with gratitude and its anniversary kept by ages yet unborn. Every heart seemed to rejoice in the occasion. I hope it will still continue to increase and in a short time produce an abundant harvest.*

Well, it was auspicious. In the halcyon days of wheat growing in the Waikato two decades later, Māori had many ploughs drawn by pairs of horses. Before they could afford horses or European-made ploughs they had made their own copies entirely of wood, which were drawn by men. As an old man told writer James Cowan, 'Ko te tangata te hoiho tuatahi' (Man was the first horse.)

Although the early missionaries may have lacked farming skill, one immigrant soon arrived who turned the Waimate mission property into the first fully mixed farm after the English model, with both crops and livestock. He was Richard Davis, from Dorset, and in 1831 he took over the 250 acres, bought from local Māori. He was a serious and efficient farmer who made an immediate success of the mission operation, which declined after he moved on.

Butler didn't stay long in New Zealand, but he was an agent of significant change. He had the previous year been a signatory, with fellow missionary Thomas Kendall, to the deed which recorded the first land sale in New Zealand. It was 4 November 1819 and the Church Missionary Society paid forty-eight 'falling axes' for 13,000 acres of land at Kerikeri from 'Shunghee Heeka' (Hongi Hika). Butler and Kendall had started the land rush.

The plough was here to stay. Double-tine versions began to appear during the 1870s and farmers graduated to three and sometimes four blades as soon as they had the bullock or horse power to do so. The first tractors appeared about 1910, without making a great impact, sometimes because of unreliability. First World War research and experiment with vehicles saw a leap in their development and from the 1920s their use became widespread, especially among those making a living from cropping.

The extra horsepower meant more tines and thus more furrows at a run. The tractor and plough were also designed so the implement could be controlled by the driver without dismounting from the vehicle. The tractor thus greatly extended the amount of land able to be ploughed in a day by one man.

But all the action wasn't in Northland. In his history of the East Coast region, *Challenge and Response*, W. H. Oliver wrote: 'The economy of the Maoris changed as they began to produce for a market. First it was potatoes and pork for visiting ships, then those foodstuffs and flax for visiting traders. The missionaries, beginning in the later 1830s, brought, among other things, improved agricultural techniques. The Maoris next learned to build and run European-type vessels.'

Oliver says the wheat boom of the 1850s, the Māori-operated export trade to Auckland from the East Coast, and the quick realisation that the land was an economic asset, all derived indirectly from Hongi and the rampaging Ngāpuhi: 'Christianity reached the region ahead of its regular apostles, carried by Maoris enslaved by the Ngapuhi, and freed after they had accepted the new religion. These unofficial new agents first spread the new ways; the missionaries came to take over a going concern.'

By the early 1840s, missionary William Williams had introduced wheat growing to Waiapu and Māori were growing and milling the grain and exporting it to Auckland.

Gisborne historian Joseph Mackay says that well before wheat was cultivated in Poverty Bay maize had been introduced and was being extensively grown by the 1830s. In the Waikato, missionary John Morgan was involved in the brief flowering of Māori agriculture during the two decades before the civil war of the 1860s, after which Pākehā snatched much of the best agricultural land for their encroaching compatriots and destroyed the sanguineness of Māori.

By all accounts, and there were many, the era of peaceful progress and industry among Waikato and Ngāti Maniapoto was from about 1845 to 1860. Then the Taranaki War, the forerunner of the Waikato campaign, interrupted the new and profitable era of productivity, especially wheat growing and flour milling. James Cowan, a writer brought up on a Waikato farm on land confiscated after the civil war, said the heavy work and the harvesting on Māori farms were done by means of an ohu (working bee). 'Waikato's surplus was taken to Auckland in large canoes,' he wrote. 'This consisted chiefly of wheat, flour, maize, pigs, fruit and muka (dressed flax). After each harvest the Waipa and Waikato rivers were busy with this canoe transport going down fully laden to the market and returning with goods purchased in Auckland. The route was via Waiuku, which was then a very busy place, and Onehunga.'

Wheat had been first introduced to New Zealand by de Surville of the *St Jean Baptiste* the same year Cook first arrived. He presented the natives with some wheat and other seeds and tried to explain what they had to do to cultivate them. Then Marsden gave Ruatara some seed at Parramatta and he brought it back to the Bay of Islands trumpeting that this was the plant from which the Pākehā made his biscuits. The wheat grew strongly, but at first some Māori, unfamiliar with grain, dug it up as it began to dry to find the food at the root. Ruatara and Hongi allowed the grain to ripen, threshed it, milled it and made a cake to convince their countrymen. That was in 1813. The wheat strain was of Indian origin via New South Wales.

Marsden gave Ruatara some seed at Parramatta and he brought it back to the Bay of Islands trumpeting that this was the plant from which the Pākehā made his biscuits.

The wet and humid summers would always prove a problem for wheat growing in the north and quite early it was discovered it did better on the drier eastern plains of the South Island. A settler, according to Franz Hilgendorf's *Wheat in New Zealand*, reported that in the 1840s, Māori on Banks Peninsula were familiar with wheat: 'When digging potatoes they sowed the grain as they dug, and in turning over the soil covered the grain. In this way they would sow as much grain as was covered by a day's digging. The crop, of course, did not ripen evenly but as they cut it with a hook that didn't matter. They cut it as it ripened.'

According to John Morgan, so much wheat was grown at Rangiaowhia, near Te Awamutu, by the summer of 1845–46 that he arranged the construction of a water-powered flour mill. Māori were at first only mildly interested but by the time the mill was finished and functioning, in 1848, the benefits were clear and many hapū around the upper Waikato area wanted mills of their own. Before the turn of the decade, there were 450 acres of wheat supporting the mill at Rangiaowhia.

'I had also introduced barley and oats,' wrote Morgan. 'Many of the people at various villages are now forming orchards, and they possess many hundreds of trees budded or grafted by themselves, consisting

'The beautiful, richly-cultivated country around Rangiaowhia and Otawhao lay spread out before us like a map. I counted ten small lakes and ponds scattered about the plains. The church steeples of three places were seen rising among orchards and fields. Verily I could hardly realise that I was in the interior of New Zealand.'

of peach, pear, plum, quince and almond; and also gooseberry bushes in abundance. For flowers or ornamental trees they have no taste; as they do not bear fruit, it is, in their opinion, loss of time to cultivate them.' The peach trees in particular seemed to thrive. The fruit was sliced and strung up in the sun and wind to dry for use in winter pies and tarts. It was so abundant it was also used to fatten pigs and even feed horses and cattle.

Austrian geologist Dr Ferdinand von Hochstetter wrote of a visit to the upper Waikato in 1859: 'The beautiful, richly-cultivated country around Rangiaowhia and Otawhao [Te Awamutu] lay spread out before us like a map. I counted ten small lakes and ponds scattered about the plains. The church steeples of three places were seen rising among orchards and fields. Verily I could hardly realise that I was in the interior of New Zealand.'

But for bucolic bliss, nothing beats the description of a Māori farming area in the Waikato in the early 1850s by Lady Martin, wife of the first Chief Justice:

> *Our path lay across a wide plain, and our eyes were gladdened on all sides by sights of peaceful industry. For miles we saw one great wheat field. The blade was just showing, of a vivid green, and all along the way, on either side, were peach trees in full blossom. Carts were driven to and from the mill by their native owners, the women sat under the trees sewing flour bags; fat, healthy children and babies swarmed around, presenting a floury appearance. ... We little dreamed that in ten years the peaceful industry of the whole district would cease and the land become a desert through our unhappy war.*

Another more fact-rich description of Māori farming came from a member of the Legislative Council, William Swainson, who wrote that a Māori population of about eight thousand in the Bay of Plenty, Taupō and Rotorua, in 1857

> *had upwards of three thousand acres of land in wheat, three thousand acres in potatoes, nearly two thousand acres in maize, and upwards of a thousand acres planted with kumara. They owned nearly a hundred horses, two hundred head of cattle and five thousand pigs, four water-mills and ninety-six ploughs. They were also the owners of forty-three small coasting vessels, averaging twenty tons each, and upwards of nine hundred canoes.*

Swainson, a numbers man, reported that in 1858 at the port of Auckland (Waitematā Harbour) fifty-three small vessels were registered as being in Māori ownership, and the annual total of canoes entering the harbour was more than 1700. Earlier, on 7 March 1848, Governor Sir George Grey had conveyed in flat, rather patronising prose in a letter to the Secretary of State for the Colonies, Earl Grey:

> *It will be interesting to your Lordship to be informed that during my journey through the extensive and fertile districts of the Waikato and Waipa I was both surprised and gratified at the rapid advances in civilisation which the natives of that part of New Zealand have made during the last two years. Two flour mills have already been constructed at their sole cost and another water-mill is in the course of erection.*
>
> *The natives of that district grow wheat extensively: at one place alone the estimated extent of land under wheat is a thousand acres. They have also good orchards with fruit trees of the best kind, grafted and budded by themselves. They have extensive cultivations of Indian corn, potatoes, etc., and they have acquired a considerable number of horned stock. Altogether I have never seen a more thriving or contented population in any part of the world.*

An anonymous writer in the *Maori Messenger* the following year described 'the singular energy and ability' of the Māori in competing with Europeans in farming and other industries as 'by no means one of the least remarkable events of the age we live in'.

Auckland fed and flourished on the food from Māori canoes, which returned to their villages with implements, clothing and much sought-after European novelties. Māori proved shrewd negotiators but did not understand the cyclical nature of capitalist economics. An agricultural depression at the end of the 1850s saw their affluence shrivel somewhat. They had come into direct confrontation with the Pākehā both as farmer and provider. In 1844, Governor FitzRoy wrote: 'The natives are well inclined to labour for very small remuneration.' He saw this as an advantage, but others saw it differently.

Their success as farmers contributed to their undoing. As early as 1845, a Dr Martin, who spent a short time in New Zealand as a journalist, wrote that Māori produced food so cheaply they would ruin the white settlers' market. The settlers had their own place for Māori in the scheme of things, and it was certainly not that of a competitor. Māori were expected rather to be the willing accomplice in converting their forests into the white man's farms.

Keith Sinclair in his *History of New Zealand* wrote: 'In agriculture the Maoris, who competed directly with the settlers, enjoyed considerable advantages. They owned most of the good land in the North Island. They farmed communally, and thus had no labour costs. Until the 1860s they probably produced the greater part of the food crops consumed in New Zealand or exported to Australia.'

British settlers had other grumbles. As they once had with their kūmara gardens, Māori farmers tended to exhaust land with their crops, then abandon it and move on to fresh soil; and their fencing was not something they dedicated much time to and their livestock, particularly their pigs, too often strayed onto others' land.

But it was the combination of depressed prices in New Zealand and Australia and then the Taranaki and Waikato wars that brought to an end Māori adaptation to temperate-zone farming in the context of a market economy — an adaptation that was an extraordinary feat measured against the performance of other 'New World' races facing the colonising onslaught of European discovery and expansion. Once the civil wars in Waikato and Taranaki were over, the best land was wrested from them for their 'rebellion'. The confident pride with which Māori had farmed in Northland, the Waikato, Bay of Plenty,

King Country, Taranaki and the Horowhenua district in the 1840s and 1850s and fed the Pākehā settlers was long neglected by New Zealand historians.

The tone of conventional history was for so long that the land was the richer for being taken over by Pākehā because Māori would leave it idle at worst or farm it for subsistence at best. But Māori mixed farmers had adapted brilliantly in many places until the pressure from politicians, speculators and Pākehā settlers began forcing them off the last of their land that was most suitable for arable and pastoral farming.

Māori fought back too late, outnumbered and under-equipped against a better organised foe, and depressed by a growing understanding of the cynical, acquisitive forces ranged against them. In selling much of their land, Māori had succumbed to the blandishments of some missionaries and of politicians. Finally, when the Pākehā stratagems would not work any more, military force was applied.

G.T. Alley and D.O.W. Hall wrote in 1940:

> *[T]he British army officers engaged in the campaigns (the British troops were withdrawn gradually after the subjugation of the Waikato) were wont to accuse the colonists of prolonging the wars for the sake of their pockets. Certainly the market for fat stock was vastly improved by army purchases. The northern settlers benefited in another way. The destruction of the military power of the Maori meant their elimination as competitors in food production, a result partly due to the voluntary retirement of the Maoris after the wars.*

A sad image remains: Rewi Maniapoto's Last Stand at Ōrākau was fought in a peach grove where wheat fields fifteen years before had held such promise for a Māori farming future.

## CHAPTER THREE

# The Evergreen Seasons

European settlers arriving in New Zealand found a lanky land running from south-west to north-east for 1600 kilometres, warmly temperate around the coasts but just wide and high enough in the central regions of both islands to produce something approaching a continental climate, with cold, dry winters and hot, dry summers. But mostly the evergreen seasons move from cool and moist to warm and moist and back again without the joy of a Hey presto! spring or the gloom of a cloistral winter.

Original Māori immigrants would have found the climate harsh in contrast to what they were used to, but for the Europeans it was benign. In almost all the farmed areas rainfall varies between 75 and 175 centimetres from showers on something like 160 days a year, with between 1800 and 2400 hours of sunshine.

The frequent alternation of rain and sunshine distinguishes the New Zealand climate from other temperate-zone conditions round the world. The prevailing wind is north-westerly — and there is a lot of it.

The earliest European settlers soon discovered the advantage such a climate bestowed on their type of farmer — that in most regions pastoral animals could be left outside on grass that grew pretty well year-round, without the need for expensive sheltering buildings. Shelter-belts of trees did the job.

A relief map of the country is a busy picture. A mountain axis runs from Fiordland up the western side of the South Island, through the south-west tip of the North Island to East Cape. Only about a quarter of the 26.9 million hectares are under the 100 metres contour. Broadly, half the country is mountainous, about a quarter of it hilly and the rest rolling or flat.

The west coast of the South Island and a strip along the south-east coast, and almost all of the North Island, were covered in heavy bush when the first Europeans arrived. The type of forest varied with altitude and geographic situation but fell into two categories: rainforest in the North Island, and some parts of the South Island such as Westland and Southland, generally at lower altitudes, with tall evergreen trees as a top storey, other lesser trees of many species and tall ferns in the middle, and a heavy latticework of climbers running up from luxuriant undergrowth on the floor. It was difficult forest to penetrate, let alone raze and develop into pasture.

Beech forest, generally confined to the higher altitudes, running along the foothills, crouched against the mountain spine of both islands.

The flat and rolling country of the South Island is clear-cut enough with a pocket at Nelson, a small plain in Marlborough, a strip along the coast of Westland, the big stretch of the Canterbury Plain, and on down through eastern Otago and round the bottom of the island into the plain of central Southland.

The high country that is farmed runs up the eastern side of the Southern Alps with grazing almost to the snowline in summer; with hill country ridges extending to the coast in North Canterbury, Otago and Southland.

In Nelson and Marlborough, the soils are recent alluvials in yellow-grey and yellow-brown earths. Quite early they were supporting market gardens, orchards and specialist crops as well as sheep and a few dairy cows. Large herds of beef cattle swarmed the high, cold rangeland of inland Marlborough.

> The west coast of the South Island and a strip along the south-east coast, and almost all of the North Island, were covered in heavy bush when the first Europeans arrived. The type of forest varied with altitude and geographic situation but it was difficult to penetrate, let alone raze and develop into pasture.

Nelson is a kind of temperate Eden with an average of 2500-plus hours of sunshine each year and between fifty and sixty inches of rain which, like mercy, is not strained and droppeth gently over an average of 120 days. This salubrious region was, ironically, the source of major ingredients of the country's cigarette and liquor production. Nelson grew all New Zealand's hops and tobacco — until tobacco was phased out during the 1990s. It continues to grow all the hops.

Grey podzols with recent swamp soils on alluvial flats support the farming on the narrow West Coast plain, which has become increasingly important as a dairying area, particularly to the south. Canterbury soils range from stony gravel to fine silts with high natural fertility, and there are extensive deposits of gravel and silt on the Southland Plain. Otago's back country has yellow-brown earths on the hills and yellow-grey earths, often stony, in the basins.

By the turn of the twentieth century stock had been carried for forty years on accessible areas of easily developed pastoral land carrying tussock on the eastern side of the South Island, the Wairarapa and southern Hawke's Bay. The only other tussock country was less accessible — a pocket in Poverty Bay and a large open range area east and south of Lake Taupō which, it turned out, had a cobalt deficiency that affected stock health. But as the century turned the signs were there — that in the twentieth, despite the heavy counterpane of forest, the North Island would wrest from the South the economic superiority its agriculture had enjoyed in the nineteenth.

Nelson is a kind of temperate Eden with an average of 2500-plus hours of sunshine each year – this salubrious region was, ironically, the source of major ingredients of the country's cigarette and liquor production. Nelson grew all New Zealand's hops and tobacco — until tobacco was phased out during the 1990s. It continues to grow all the hops.

The major farming regions of flat and undulating land in the North Island are:

- *The Waikato-Thames-Hauraki Plains*, which are composed mainly of volcanic sediments derived from the central plateau. The plains of the North Island are usually composed of finer sediments than those of the South Island. A considerable portion of the area was formerly swamp, but drained, it is now good farming land. The Waikato-Thames-Hauraki Plains are separated by a low range from the small, disconnected plains of the Bay of Plenty.
- *The Poverty Bay Flats* are basically the valley plain of the Waipāoa River. North and south of this plain, discontinuous patches of flat land occur in the valleys of minor streams and along coastal strips.
- *The Heretaunga Plain* is the most extensive area of flat land on the east coast. It is roughly triangular and is formed of alluvial sediments deposited by the rivers which cross the plain. After the earthquake of 1931 land was reclaimed from the Ahuriri Lagoon, near Napier.
- *The Taranaki Plain* is an elevated plain ending at the sea in cliffs. Its surface is broken by the courses of rivers which emerge from the hill country. Thus there are small areas of river-terrace plains lying at a lower level than the general elevated surface of the country.
- *The Manawatū-Horowhenua Coastal Plain* stretches from the Manawatū Gorge to the coast. From the gorge to Palmerston North the plain slopes steeply, but between the town and the sea the slope is more gradual and the river follows a winding course across it. This part of the plain is separated from the sea by a line of sand dunes. The rainfall on the coastal belt is sufficient for dairy farming on the light sandy soil. Southward the coastal plain becomes narrower and is then known as the *Horowhenua Coastal Plain*, as it is largely contained in Horowhenua District.

This plain consists of a series of terraces along the foot of the high country, while on its outer edge it is fringed by sand dunes and swampy hollows. These hollows, which are numerous in the vicinity of Levin, were formerly covered with flax. Foxton was once the centre of the flax industry in the nineteenth and early twentieth centuries.

- *The Wairarapa Valley* runs parallel to the high country and separates it from the coastal hill country. In its upper portion the plain slopes steeply, but southward it widens and the slope is more gradual, and much of the land is good dairying country.

In the North Island hill country, laboriously farmed with sheep, the major regions are:

- *The North Auckland Peninsula*, which is complex in character. The chief areas of flat land are those along the rivers in the coastal areas. The region is mainly a soft rock upland of low elevation, generally below 200 metres, but with several peaks rising to 360 metres. Outcrops of resistant rocks, mainly volcanic, form outstanding and resistant masses. The eastern coastline of the peninsula is often irregular, with deep inlets, extensive estuaries and mud flats, while on the west coast there are long stretches of sand dunes, notably along the Ninety Mile Beach in the far north.
- *The Central Volcanic Plateau* occupies the greater portion of the middle of the North Island. The main systems of river drainage radiate outward. The plateau is built of volcanic material, its surface being relatively even, but often broken by volcanic cones and irregular ridges. Large areas have been used for the establishment of exotic forests and farming has been successfully established.
- *Inland Whanganui-Taranaki Hill Country* fringes the western boundary of the volcanic plateau from which it stretches to the volcanic cone of Mt Taranaki/Egmont. The hill country surrounding it is composed of soft rocks. The highest ridges are those further inland and overlooking the volcanic plateau, rising in the east to 900 metres. From the highest ridges there is an even slope of the land towards the coast. The whole plateau has been intricately and deeply dissected by streams.

  The ridges between the stream valleys have sharp crests and steep sides. This area of weak rocks and steep hill slopes is also one of high rainfall, where formerly the soil was protected by a dense forest cover. Much of this type of land has been cleared and used for the grazing of stock; today some areas are reverting to second growth, while there is some soil erosion on the steeper slopes.

- *The East Coast Ranges and Hill Country* extend from Cape Palliser to East Cape, about 550 kilometres. Throughout the southern section the ranges are narrower than they are to the north of the point where they are cut by the Manawatū River. From this river gorge they widen out, the Kaimanawa Range extending westward towards the volcanic plateau as a wide belt of rugged mountain country. On the eastern side of the main range there is an extensive belt of hill country stretching to the coast both to the north and south of Hawke Bay. The northern ridges of this soft-rock hill country lose elevation seaward.

  The slopes facing inland are steep, while larger and more gradual slopes face the coast. Few large rivers run through these valleys and it is only in the lower valleys that there are any areas of flat land. To the south of Hawke Bay the main ranges are separated from the coastal hill country by low rolling land or by wide gravel valleys.

North Island soils have been strongly influenced by the fallout of volcanic ash, with surface deposits found as far south as Wellington and as far north as Kaikohe. But the area dominated by volcanic ash soils is the centre of the island, from Whanganui to the Waikato Heads on the west and from Napier to Whitianga on the other coast. In the centre and along the east coast the predominantly yellow-brown pumice soils are low in elements, notably phosphorus, potassium, magnesium, cobalt, copper, iron and boron. To the north, west and south the older volcanic ash has formed yellow-brown loams, less deficient in trace elements.

That is how an agriculturist sees the land that is New Zealand, but it does not fully describe the difference between the two main islands — that in the South there is a different scale of distance and size, longer flats and higher mountains, with bold, faraway horizons that allow a special kind of reflective patience to overlay a person's thinking. The southern pioneering farmers had time and space and few obstacles to overcome, so their development was predictable and sure. They are steady and tough and by temperament conservative, and as the end of the nineteenth century approached, they seemed to have run out of drive. And debt was an encroaching problem.

In the North Island the landscape is newer, smaller in scale, with a close and shifting horizon bringing surprise with every twist of a journey. Northern settlers found this broken countryside with its endless walls of trees a dogged adversary that gave them little time to think during the thirty years from 1880, only time to slog at the forest face, cutting and burning and sowing grass in warm ashes. Turn their back on the bush and it writhed remorselessely up through the grass again.

They took the bush on and won. But it was a long hard fight.

Chapter Four

# The Pastoralists Arrive

The first post-European New Zealand farming on a large scale was wool growing, an export-orientated enterprise that brought this country its first taste of affluence. It developed in the south of the North Island at the time the zenith was approaching in the north for the Māori-dominated mixed agriculture that fed the earliest settlers and opened the export trade in food.

The first pastoralists were from Wellington — the least likely town, locked as it was by mountains and sea from the grassed valley flats or undulating tussock land pastoralists needed for sheep. These men got the idea, and then the sheep, from Australia where fortunes were being made from Merino wool.

Wellington juts its jaw out over the edge of the South Island and is often hit by a southerly uppercut brought up from the floor of the world. The original idea was to build the city in the Hutt Valley, on the present site of Petone. The first settlers in what was called Britannia were flooded out in 1841 by the Hutt River and moved onto the Te Aro flat and the hills of Wellington. This flood disqualified the valley as a possible sheep run, for the time being anyway. Soon, the New Zealand Company's close-order, tidy style of settlement around Port Nicholson (Wellington Harbour) began to fail for want of suitable land to farm; and local Māori disputed ownership of what little there was.

Life was hard for Britain's working class and lower middle class at the time, and especially for rural workers. The steadily increasing enclosing of common land denied them their subsistence farming. The Industrial Revolution was driving them into the towns and cities where poverty was worsened by the immense population growth during the nineteenth century.

A driving force behind the enclosures in England and Scotland was an agricultural revolution that began in the late eighteenth century. Animal and plant breeding and planting techniques were improving and mechanisation was increasing, reducing the demand for labour and providing an incentive for wealthy landowners to fence off their own land and then common land. They did so with little thought for the chaos that would ensue. Unemployment led to pauperism and then to spreading lawlessness in the densely packed cities.

Edward Gibbon Wakefield and his New Zealand Company associates had theorised the development of a colony, using the village structure of self-sufficiency gradually spreading outwards, selling land to settlers and using the payments to bring out more migrants and to buy more land.

The scheme was not working in Wellington, which was by 1842 an unhappy town snuggled in bush that made the land difficult and expensive to get into production. As Joseph Masters, the man who helped found Masterton, wrote in a memoir (published by The Wairarapa Archive in 2005): 'The only land available around Wellington at this time [the late 1840s] was the wooded hills around Wellington, totally unfit for agricultural purposes.'

The small farmer concept was vested with much Protestant virtue. It was considered the right sort of thing to do as much in moral as economic terms. G.T. Alley and D.O.W. Hall noted in their Centennial Survey history: 'In Otago the *News*, within a short time of the beginning of settlement, repudiated the idea of agriculture altogether, advocating the development of pastoral farming as the only road to prosperity, an advocacy that so angered the citizens of Dunedin that the paper came to an end late in 1850 for lack of influential support.'

It became obvious to one group of Englishmen in Wellington, ensnared only briefly by the New Zealand Company dream, that a better way for a man with capital to get into fast money was to find grassland and graze sheep. If arable farming had about it a work ethic air of wholesomeness for the working class and lower middle class, sheep farming was glaringly a business enterprise that appealed to the instincts of the upper-middle-class Englishman. So these men, some of them related, formed partnerships to invade the Wairarapa Valley.

One of them, a future Premier, Frederick Weld, according to his biographer, said: 'I never had any taste for pottering about with little bush cultivations, an occupation more suitable to labourers with large families.'

Future Premier Frederick Weld, according to his biographer, said: 'I never had any taste for pottering about with little bush cultivations, an occupation more suitable to labourers with large families.'

Almost all the Wairarapa pioneers' capital went into sheep; initially the land was the least of their financial problems. According to R.D. Hill in *Pastoralism in the Wairarapa*, there were 'some seven hundred and eighty members of the Ngāti Kahungunu tribe in the Wairarapa when the squatters arrived'. They were pleased to see Pākehā because they had been decimated by invading tribes and knew the presence of the settlers would give them some protection.

The valley floor and low hills were covered with bush and scrub, and grass and fern, with some long stretches of swampland especially around the lakes. It was the kind of flat, open land — comparatively rare in New Zealand and particularly in the North Island — that suited large flocks of sheep. The trick was to get them into what was then a valley sealed off by mountains and the sea. The southern end was subject to the same southerly king-hit that afflicted Wellington, but the ranges to the west kept all but the most southerly part fairly dry.

The first white men to peep into the valley were peripatetic Austrian naturalist Ernst Dieffenbach and the surveyor-painter Charles Heaphy who 'had a fine view of Palliser Bay and the Wairarapa Lakes' from a southern high point in the Remutaka (Rimutaka) Range while on an 1839 expedition in search of the then rare and now extinct huia bird.

The first European to wind his way around the south cost, over the Mukamuka rocks and into Palliser Bay was William Deans early in

1840. He pondered the prospect of settling in the Wairarapa Valley, but then decided to go south. Later he and his brother, John, settled on the then lone and level land of what became Canterbury, and set up the famous Deans Farm at Riccarton, an island of dry flat land in what was mostly swamp behind the Port Hills.

Two expeditions, in 1841 and 1842, really gave the Wairarapa a going over for the first time by Europeans. First, New Zealand Company surveyor Robert Stokes walked in with a companion, J.W. Child, and two Māori guides. Another two New Zealand Company surveyors, Charles Kettle and Charles Brees, were next and after an arduous month, which they probably wouldn't have survived but for food and guidance from local Māori, they walked out into the Hutt Valley over a route similar to that of today's Remutaka highway. Brees reported of this pristine pastureland that it had 'an abundance of prairie grass and fernland for pastoral purposes; a sufficiency of timber for building and fuel', with 'quite sufficient land for arable purposes to suit settlers'.

In 1843, a group of would-be squatters did a deal with the Māori inhabitants. Frederick Weld, Charles Clifford, William Vavasour and Henry Petre (all from the same prominent English Catholic family) leased about one square mile of grass and scrubland and a huge area of hill country on the drier eastern side of the valley for £12 a year.

Others were involved with separate deals, notably Charles Bidwill and William Swainson junior. These two were the first to use the rough coastal route, carrying the sheep over the Mukamuka rocks one at a time, 350 of them, mostly in-lamb Merino ewes. It was what Bidwill called, 'A work of time but which was accompanied with less trouble than we expected and without any loss.'

With autumn closing into winter, Bidwill and Swainson had passed the Weld party with their 600 Merinos on their way round the wind-lashed, sea-thrashed southern beaches. This second group lost some of their sheep into the sea as the weather worsened around the Mukamuka rocks (which were later lifted high above the shoreline by the 1856 earthquake). Thus the first two sheep stations in New Zealand were on grazing leaseholds held by Weld, Clifford and Petre at Wharekaka Station, and Charles Bidwill at Pihautea.

A combination of the squatters' inexperience as pastoralists and the marginal adaptability of the Merinos to the heavy lush land made the start a tortured one. The Māori landlords moved closer and with their help and the hunting of game birds the newcomers fed well enough through the first year until they could get their own gardens in shape, but they wondered why their sheep did not thrive as well as expected. Then experienced Scottish and North of England shepherds came to the rescue, first by moving the sheep off the lower land where Merinos proved susceptible to foot problems.

After seven years as a pastoralist, Weld was knowledgeable enough to write a best-selling pamphlet, *Hints to Intending Sheep-Farmers in New Zealand*. It ran to at least three editions. Scab (a form of allergic dermatitis caused by the highly parasitic scab mite *Psoroptes ovis*) remained a problem, though, and labour was not easily available. The valley was especially good land for cattle and before long some of the increasing number of settlers were complaining that too many of them were involved in the cattle trade for it to be profitable. Wool prices were climbing all the time, though, and profits were solid.

Within two years twelve stations spread across the Wairarapa. Weld, Clifford and Petre established Wharekaka Station on the shores of Palliser Bay. By 1845 they had 900 ewes and were wintering sheep for newcomers. But Petre sold his share to his partners and Weld began to find life hard in the isolation of the Wairarapa. He and Clifford decided to expand their business by buying a station in the Wairau Valley in Marlborough. Vavasour also went south.

Weld and Clifford bought land at Flaxbourne (where the town of Seddon is now) and Weld wrote: 'I shall, with my yacht, be much nearer the town [of Wellington] than I am now. I shall have no rivers to ford, sometimes breast-high, no rocks to climb at high tide on the beach, or to sleep out in the rain all night. Nor shall I have any anchoring off lee shores in open boats, or swamping in the surf, of which I have had enough to last me for years.' The first track was cut across the Remutaka Range in 1847, and a road was completed for wheeled traffic in 1859. Until then settlers and their wool travelled to Wellington by sea.

Towards the end of the decade, Weld and Clifford had 11,000 sheep at Flaxbourne, and by the mid-1850s had more or less abandoned

Wharekaka. Weld had a thirty-six-foot boat, rigged like a cutter, which took them across Cook Strait to Wellington and back.

Francis Dillon Bell, a Wakefield relative, was sent into the Wairarapa Valley in 1846 to buy 250,000 acres for a prospective Church of England settlement, presumably expecting incumbent pastoralists to pick up their sheep and disappear before the Lord's shepherds. Dillon Bell had written support from Governor Grey but made no progress with the Māori owners and got short shrift from the squatters. God and Grey willing, the Anglicans decided to go instead to Canterbury once the Free Church of Scotland settlers had rejected it in favour of Otago.

The Wairarapa remained under lease and it wasn't until the mid-1850s that the squatters managed to get title to their properties. Meanwhile, the wool grew fast and the profits faster and by 1850 many of the original pastoralists had quadrupled their investment. Weld, by 1852, had netted £4000, a heady sum in those days. Pastoralists stretched their enterprise into Hawke's Bay, which was almost entirely stocked from the Wairarapa, as well as Marlborough, Nelson and Canterbury. The squatocracy had arrived, educated men who had capital and the kind of confidence that Victorian Englishmen wore as their birthright. They called themselves 'colonists' and the others 'emigrants', or 'settlers'.

By 1855, sheep numbers in Marlborough and Hawke's Bay had edged past the Wairarapa. By 1861, Wellington (mainly the Wairarapa) had 250,000 sheep, fewer than the 312,000 in Hawke's Bay, and 368,000 in Marlborough. Further south, the Canterbury flock had more than trebled in six years to 900,000 and in the same six-year period sheep numbers in Otago had, astonishingly, grown from 75,000 to 617,000.

From then on the South Island was easily the predominant producer of wool and sheep meat. The returns were spectacular through the 1860s. Wool exports earned £444,000 in 1860, £1,704,000 ten years later, £3,398,000 in 1875 but down to £3,169,000 in 1880 as prices slumped at the beginning of the country's longest economic recession. They did not pick up again for more than a decade. Sheep farmers' price woes were compounded by the spread of scab and the waves of rabbits eating out pastures.

Even more spectacular in the 1860s and 1870s was the increase in the value of grain exports from £18,000 in 1860 to nearly a million pounds twenty years later. The true wheat bonanza was the five years between 1875 and 1880 when the returns from grain (mostly wheat) quadrupled from £240,000 to £921,000. This explosion of grain exports was made possible by increasing farm mechanisation and the creeping tentacles of the railway system. It was also mainly a South Island enterprise.

Local histories have described how hard it was for settler families in the very early days of the Wellington-Wairarapa region. A passage in *Brave Days*, written for the New Zealand Centenary in 1940 by members of the Women's Division of the New Zealand Farmers' Union (predecessor of Federated Farmers) says:

> *In those early days journeys over the Rimutakas were memorable undertakings. Of the women, who with young children accompanied their menfolk on these treks, too much praise cannot be given … undaunted by the rough roads, flooded rivers, storms, privations and dangers of all kinds.*
>
> *In spite of their own trials, which in the nature of things must have needed more endurance than those of the men, they were always ready to encourage and inspire. It is therefore a strange thing that, in almost all the published accounts of those days, given by descendants of these settlers, little tribute falls to the lot of these noble women.*

Stories told of women having babies unattended in storm-isolated farmhouses, about families waiting for days to cross rain-raged rivers, about people facing Māori bands lashing back at land-grabbers. The isolation was very real and it was the ability of them and their families to endure it and to produce wool by the tonne that put the mewling New Zealand economy firmly on the sheep's back. Wool started to earn New Zealand a stronger living in the world, the first non-extractive, renewable farm product to do so on any scale.

Pastoral development rushed ahead in the south because of the vast areas of tussock land. The stations remained key economic units — despite a slump in wool prices from the late 1870s — until the

Liberal Party's Minister of Lands, John McKenzie, set about cutting them up in the 1890s in a bid to beat landlordism.

As the sheep farmers with big flocks tended to move to other more suitable regions, the character of farming in the Wairarapa began to change. The first New Zealand Company settlers in Wellington, Nelson, New Plymouth and then Wanganui (a spillover from Wellington when the land shortage became critical) found the land they had been promised on the other side of the world was largely unavailable. The twenty million acres the company claimed to have bought in the central region of the country were cut back by the government to less than 300,000 after the Treaty of Waitangi was signed about the time these early settlers arrived.

Arthur Thomson in his 1859 *The Story of New Zealand* wrote that settlers who had not got possession of lands as they expected were 'frittering away their lives in idle pastimes, while torpor and drinking had taken possession of a few'. But as well as these wasters there were those with very limited access to capital who seriously wanted enough land for subsistence farming.

In Britain, land ownership bestowed status and was insurance against pauperism. While many Wellington settlers grumbled and succumbed to torpor, Joseph Masters decided to do something about it. He was an industrious Englishman, born in 1802, who arrived in Wellington in 1842 and set up business as a cooper on the Lambton Quay beach. He and his son-in-law also began business as storekeepers and ginger beer manufacturers. His wife and daughter made straw bonnets.

Arthur Thomson in his 1859 *The Story of New Zealand* wrote that settlers who had not got possession of lands as they expected were 'frittering away their lives in idle pastimes, while torpor and drinking had taken possession of a few'.

Masters was particularly scathing about Wellington's first settlers who were frustrated by lack of land. '[They] were, with very few exceptions, very ignorant people, very few were practical farmers. Many of them were drunkards, and some worse; a number from the parishes and workhouses around London, together with a good sprinkling of Glasgow weavers. There were some good labourers from the provinces.'

After serving his apprenticeship as a cooper, Masters became one of Sir Robert Peel's 'Peelers', a London policeman. He also worked as a soldier and a jailer. He spent ten years in Tasmania before coming to New Zealand with his family. A central figure among those working-class settlers in Wellington who wanted to buy land, he began writing letters to newspapers in the early 1850s advocating small farm settlements, especially in the Wairarapa. Governor Grey was keen on the idea and the Superintendent of Wellington Province, Isaac Featherston, was also a supporter.

Masters helped form the Small Farms Association which sent him to visit the Wairarapa to choose suitable land and to persuade local Māori to sell. He claimed later that he lured Māori interest with the promise of shops like those in Wellington in new Wairarapa towns. Grey, not long before he left New Zealand for a new job in South Africa, arranged for 25,000 acres of land in the Wairarapa to be made available for sale.

A ballot was held for parcel lots of one town acre and forty acres for farming on the outskirts. Masters also organised finance. As a result, in 1854, the land on which Greytown (after Sir George) and Masterton (after Masters) were built was bought and settled. Charles Carter, a more recent immigrant also involved in the association, proved a steadier hand than the quarrelsome Masters and became the leading figure among the string of towns that quickly included, eponymously, Carterton and Featherston, established with the backing of the provincial government.

The land on which Masterton was built was originally claimed by a squatter, William Donald, who arrived on the heels of Weld and Clifford in the early 1840s, and whose descendants still farm sheep in the Wairarapa. William Donald's son, Donald Donald, invented a wool press which was essential equipment in many New Zealand and Australian wool sheds for 150 years. The manufacturing firm of Donald and Sons, formed in 1900, made the presses and other Donald Donald inventions such as wire-strainers, saw benches and tractor-powered post-drivers.

The population of the towns doubled before the end of the 1850s, and Masterton was and still is the main town. The farmers who did best and survived were those who aggregated land by buying from those who gave up. Masters became unpopular not only because he bought up much land but also because he ran sheep on it contrary to the spirit of the small farms. But the Wairarapa remained predominantly sheep country, although the small farmers sold dairy products, grain and fruit and vegetables on the Wellington market. In the beginning, many of the subsistence farmers supplemented their income with outside work, often on road making and maintenance or on railway maintenance, and sometimes on the sheep stations.

In modern times, the Wairarapa has again attracted small farmers — of a very different kind from the people who followed Masters and Carter into the then inaccessible hinterland of Wellington. As the road over the Remutaka Range was modernised and particularly after a railway tunnel was cut through from the top of the Hutt Valley in 1955, many city-dwellers moved out to distance themselves from the hustle and bustle as commuters; or to retire; or in search of an alternative lifestyle; or as agricultural entrepreneurs to live off the land in new, diverse and spirited ways; or perhaps to grow grapes and make wine at Martinborough named, in the Wairarapa tradition, after its founder, Irish-born politician, John Martin.

CHAPTER FIVE

# Settling the Middle (South) Island

The second half of the nineteenth century belonged to the South Island — or Middle Island as it was called early on, investing Stewart Island with a geographical significance it could not sustain — and this was because of wool, wheat and gold.

The Free Kirk Scots had first choice of a settlement site south of Nelson. Lyttelton, Port Cooper then, was the front runner during the early planning. The Rev. Thomas Burns, a nephew of Robert and a stern family backlash to the philandering poet, wrote to the proposed expedition leader, William Cargill, in 1844, more than three years before they set out: 'I conclude that Port Cooper is our place of destination. I should be sorry that any change were made from that locality...'

Well, an irascible Quaker surveyor called Frederick Tuckett, entrusted with the on-the-spot choice of a site, changed their minds in favour of Port Chalmers. At first Burns suspected some dark conspiracy by Governor FitzRoy and the Church of England, but ultimately was persuaded the choice was a good one, comforted by advice that wood and water were not plentiful on the Canterbury Plain. And so the south was carved up — the Scots in Dunedin (from 1848) and the English in Canterbury (from 1850).

The New Zealand-born politician and writer William Pember Reeves wrote of the two groups in *The Long White Cloud:*

> *When the stiff-backed Free-Churchmen who were to colonise Otago, gathered on board the emigrant ship which was to take them across the seas, they opened their psalm-books.*

*Their minister, like Burns's cottar [farm labourer], 'waled a portion wi judicious care,' and the Puritans, slowly chanting on, rolled out the appeal to the God of Bethel:*

*God of our fathers, be the God*

*Of their Succeeding race!*

*Such men and women might not be amusing fellow-passengers on a four months' sea voyage — and indeed there is reason to believe they were not — but settlers made of such stuff were not likely to fail in the hard fight with nature at the far end of the earth; and they did not fail.*

*The Canterbury Pilgrims, on the other hand, bade farewell to old England by dancing at a ball. In their new home they did not renounce their love of dancing, though their ladies had sometimes to be driven in a bullock-dray to the door of the ballroom, and stories are told of young gentlemen, enthusiastic waltzers, riding on horseback to the happy scene clad in evening dress and with coat-tails carefully pinned up. But the Canterbury folk did not, on the whole, make worse settlers for not taking themselves quite so seriously as some of their neighbours.*

*The English gentleman has a fund of cheery adaptiveness which often carries him through colonial life abreast of graver competitors.*

For all Reeves' recommendation, some of the English settlers of the day took themselves with chosen-race, right-hand-of-God seriousness. An anonymous Indian Army officer, who visited New Zealand in the very early 1850s and hurried back to India to hand in his commission and return to settle in Canterbury, wrote in a magazine article:

*It was a scheme to carry out complete, to an uninhabited land, a section of English society; to open the colonies to gentlemen; to transport knowledge and learning, the arts and the sciences; to prove to the world that the energies cramped by poverty, and the crowd at home, may have scope, by emigrating, without losing all that makes life valuable; and to build up a society*

> *on that foundation which experience had taught them to be best — the Protestant religion. It was a noble scheme, and Providence, as if to second such an effort, threw into their hands a tract of land perhaps better adapted than any other for the object they had in view.*

Only the culturally blinkered could see an 'uninhabited' land.

The Canterbury Association had an even greater predilection for orderly and close-knit settlement of their province than the New Zealand Company had in other areas; and it had an even more rigid formula for the future. It wanted to establish a microcosmic England with a landed gentry, a middle-class and labourers, all prayed over and preyed upon by the Church of England, but without the poverty and ignorance that pervaded the lower levels of British society at the time. It was utopianism to plan a social and economic structure and expect to avoid the by-products of the society it was exactly modelled on.

But, as in Wellington, reality intervened, and the plan lived on as only a plan, mostly as an illusion. The association felt able initially to charge £13 an acre for land in the new settlement — one-third to be granted to education and the Anglican Church, one-third to be spent on immigration, a sixth on administration and public works, and the remaining sixth to go to the New Zealand Company for the land. A 500 per cent mark-up.

Governor Grey did not like the price much, especially the slice allocated to the Church, and the settlers liked it so little they bought only £50,000 worth in the first year, 1851, instead of the expected £1 million. What actually happened was decided, as it usually is, by a combination of ingenuity and economic opportunities discerned. Settlers were not to be confined to town by a philosophy of settlement.

Squatting started on a large scale. The sheepmen moved out quickly from Christchurch and soon great flocks were spreading over the plain. Clifford and Weld had taken sheep into the open country of Marlborough as early as 1846 and were joined in the range grazing of sheep by a number of leaders of the Nelson community. They became a major source of sheep for the Canterbury runholders in the following decade, although imports from Australia continued for a number of years.

In the 1850s, the price was still relatively high for wool and demand for increased quantities, although it wavered on occasions, remained basically strong. Run-holding held up economically until well into the 1860s. New Zealand's all-time best-known sheep farmer, Samuel Butler, arrived in January 1860 and when he left in the middle of 1864 he had doubled his investment in his Canterbury holding at Mesopotamia. Three or four years later he might not have been so lucky.

Some sheep farmers were ruined by the big snow of 1867. The reasons for wool's buoyancy were complex but certainly involved was the discovery of gold in Australia, which drew hundreds of men off the land in the 1850s in search of a fast fortune. Also a factor was the accelerated movement of the English and Scots off the land and into the cities in Britain as the industrial revolution gained pace and outstripped the agricultural revolution.

The Wairarapa pastoralists and their shepherds had been predominantly Englishmen, with a few Scottish shepherds. But from the first years of the Canterbury settlement a new influence arrived: Australian sheepmen. L.G.D. Acland wrote in *Early Canterbury Runs*: 'There had been a very bad drought in Australia in 1850 which disgusted many of the squatters there with the country. Some of them sold or abandoned their runs and came to New Zealand early in 1851 to try their luck again with what money and stock they had left. They not only brought money and stock but experience, which was most valuable to the young colony.' The squatters had 'unlimited faith in the squatting system, and a great contempt for the Canterbury Pilgrims' desire for freehold agricultural farms. The Australians were nicknamed "Prophets" or "Shagroons" and it was they who nicknamed the association's settlers the "Canterbury Pilgrims"' — with irony rather than respect.

'Shagroons' was the name that stuck. These tough and taciturn men, hats hung low on bobbing heads, almost appendages of their stringy Australian horses, drove their sheep past the less assured Canterbury Pilgrims and led the thrust on to the plains beyond.

Their macho attitude was not an affectation. They had endured the dogged dust and distances of Australia and indomitably mustered their sheep for the hard trip to New Zealand when their holdings in that other country could not be tamed. They knew all there was then to know about sheep, especially about Merinos, the predominant breed in both Australia and New Zealand at that time. They were notoriously impatient of any restraints upon grazing the unpopulated plains.

The Australians were nicknamed 'Shagroons'. These tough and taciturn men, hats hung low on bobbing heads, almost appendages of their stringy Australian horses, drove their sheep past the less assured Canterbury Pilgrims and led the thrust on to the plains beyond.

Pember Reeves wrote: 'Coming to Canterbury, Otago and Nelson, they [Australian sheepmen] taught the new settlers to look to wool and meat, rather than to oats and wheat, for profit and progress. The Australian coo-ee, the Australian buck-jumping horse, the Australian stock-whip and wide-awake hat came into New Zealand pastoral life, together with much cunning in dodging land-laws, and a sovereign contempt for small areas.' A number of Aboriginal and part-Aboriginal stockmen came out with the Australian squatters and worked for them and other early South Island runholders. The Otago historian Herries Beattie claimed that as a young man he often heard sheep called 'jumbucks'.

By temperament and background many of the shagroons were more akin to whalers and sealers or the bushmen of the North Island than to the genteel middle-class Canterbury Association settlers. There was a mutual contempt too between the agricultural men and the pastoralists, as there had been in the Wairarapa. For the Australians, sheep farming was the manly way to live on the land and the good news they discovered soon was that because of the fast-growing grass in a more benign climate, the weight of Merino fleeces averaged four pounds a head in Canterbury compared with around two and a half pounds in New South Wales. Years later

Herries Beattie, historian of Murihiku (the south of the south region), wrote: 'In pastoralism we copied Australian ways rather than those of other lands.' But he noted that the term 'squatter' was not used 'except jocularly' in the South Island and that the rest of the Australian pastoral patois disappeared quickly.

Since the 1840s an argument raged around whether the leasing of land was illegal, but in late 1851 the Canterbury Association Resident, John Godley, changed the rules and allowed unused land to be taken up by runholders in lots of 5000 to 50,000 acres, according to locality. That set off a scramble for land, the plot of so many American Westerns, told matter-of-factly by Butler in his *A First Year in Canterbury Settlement.*

The procedure, as Butler and others have described it, was to ride out and stake a claim, then apply to the land office and if your claim was first the run would be yours. The runholder then had to stock it within six months, one sheep to every twenty acres or a head of cattle to every 120 acres, and he had to pay a farthing an acre rent for the first two years, halfpenny an acre for the next two, and three farthings for the fifth and all subsequent years.

Acland wrote:

> *If he did not fulfil these conditions he forfeited the run. No term was stated to his licence, but he or anyone else could buy the freehold of all or any part of the run at any time, except that if he built a hut or made a fence or put any other improvement on the run, the improvement gave him the pre-emptive right over so many acres adjoining it. Anyone could challenge his pre-emptive rights at any time, and if the owner didn't buy the land challenged within a month the other man could, and then the runholder got nothing for his improvements.*

Gaining and retaining land through the period of squatting in Canterbury was difficult. It called for an understanding of complex legislation and a deep and ruthless cunning. It is best described at length by Randal Burdon in his *High Country.*

Livestock had been farmed in the South Island since the 1840s with small flocks of sheep, larger herds of cattle and also horses imported by whalers. The ex-whaler Johnny Jones had about 2000 sheep, 200 cattle and 100 horses at Waikouaiti in 1844, on a property called Matanaka. Banks Peninsula was settled in 1841, and the Deans brothers were well settled with stock at Riccarton by 1843. From 1850, with the spread of the runholders, sheep and cattle grazed the grass and tussock land of Otago and Canterbury.

By 1854 an estimated 100,000 sheep thrived in Canterbury and four years later more than half a million; and by 1858 a quarter of a million grazed Otago tussock. At that time, Canterbury was either deep swamp or heavy fern country along the coast, with large tracts of tall mānuka scrub behind it, backed by heavy tussock further inland.

A number of stations were first stocked with cattle and some farmers continued to breed and fatten large herds. Following the discovery of gold in Central Otago and, later, in Westland, New Zealand's population soared from 60,000 to 200,000 in the 1860s, creating a growing demand for meat; but until refrigeration became a commercial proposition in the late 1880s, cattle did not make the economic sense that sheep did. The heavy trampling of cattle and their ability to survive on more rank growth than sheep made them a sensible animal for grazing new country, but once the natural pastures were improved or English pastures patched in by burning and sowing, then sheep did better and the use of them both for complementary grazing was generally not practised in those very early days. Acland wrote that by the 1860s there were about 250 stations in Canterbury, and that perhaps twenty of them, mostly along the coast, carried cattle.

Every station had house cows and, before tracks were flattened enough for the up-tempo pace of horses, bullocks were the main draught animals — muscling inexorably along in front of a sledge or dray. Training and working the bulls was an especially tough job and bullockies became famous for their ear-blistering language. Bullocks were almost exclusively used for first-time ploughing of New Zealand land.

The Wairarapa pioneers, Clifford, Weld, Vavasour and Bidwill, the first to bring sheep into New Zealand in substantial numbers, farmed only Merinos. The breed originated in Spain and came to this country direct from Australia after being influenced as a breed by farmers in a number of other countries, notably France, Britain and South Africa. The Merino is tough, active, wiry, nervous unless up in the hills with plenty of room to move, and is susceptible to footrot on heavy ground.

Its wool is compact and fine, its meat lean and stringy. It was not the ideal sheep for the spongy valley floor of the Wairarapa or the swampy land immediately beyond the Port Hills, but up in higher country it thrived, as it did on the drier hills and plains of the east coast of the South Island, Central Otago and Hawke's Bay.

Before refrigerated shipping began the export meat trade in the 1880s, there was an enormous surplus of mutton in New Zealand. Merino carcases were boiled down for tallow but that didn't matter; their sole reason for existence then was to grow wool. Merinos were adept at fossicking for food in the often rank tussock country and were less troubled by footrot in the South Island than in the North. The sheep were driven on to new land under the eye of a shepherd and, with luck, natural boundaries to the run made the task of keeping the flock at home easier. In some cases, dogs were tied on long leads to boundary posts to bark sheep back to their owner's property.

An early task for the settlers was to replace tents with whare (huts), primitive at first, possibly wattle and daub or sod, thatched with bulrush or tussock; and then perhaps another whare fairly soon, at a strategic place, for a shepherd. The next stage would be a cob homestead, followed ultimately by timber with iron, slate or shingle roofs.

The runholder simply put his sheep on the run and directed most of his energies to keeping them there. He marked the sheep when necessary and shore them when it was convenient. With ewes and rams often running free in hard country, mustering was difficult and the times of lambing and shearing were hard to control. Lambs dropped between May and August suffered heavy losses and every year some sheep would be mustered up with two years' fleece, having been missed in the muster the year before.

Shearing was in the earliest days and remains the major production task on sheep runs large and small. It was back-breaking work with blade shears often in open paddocks. Machines became available in the first decade of last century, quite quickly replacing blade shears except in high country where sheep suffer in the cold from too close a shave. The first source of power for the shearing machines was the portable steam engine, which was replaced by internal combustion systems and then by electric power as reticulation spread through the rural districts in the 1920s and 1930s.

The shearing and farm handling of wool was where the highest levels of strength and skill were required for the efficient harvesting of wool, and it wasn't long before teams of shearers with their 'fleecies' were highly valued and highly paid, moving from run to run in the later winter or early spring, depending on the local weather. It even became a spectator sport. The Bowen brothers — Ivan and the more flamboyant Godfrey — brought about a revolution in shearing technique and turned the hard graft of rural work into entertainment for townies all over the world.

The Bowens first took black singlets into the limelight in the early 1950s with world shearing records. Godfrey, short and thickset, ideally built for the job, had refined his shearing technique since the age of sixteen and had the kind of charisma that made him a star. He elevated the job of shearing first into a dramatic competitive sport and then into international entertainment.

The brothers were two of four sons of a Hawke's Bay sheep farmer and it was on the job that they changed the basic approach to shearing and became the fastest and most proficient in the world. Godfrey broke the world record in 1953 by shearing 456 full-woolled sheep in a traditional shearers' nine-hour day. The brothers each went on to break a string of records and were first and second in the inaugural national Golden Shears contest. They were hired by the New Zealand Wool Board to demonstrate and promote the new technique, with the emphasis on both speed and quality. Wool was glamorised as a by-product of their performances.

The Bowen brothers — Ivan and the more flamboyant Godfrey — brought about a revolution in shearing technique and turned the hard graft of rural work into entertainment for townies all over the world. Godfrey broke the world record in 1953 by shearing 456 full-woolled sheep in a traditional shearers' nine-hour day.

This was of special importance to the industry because the problems besetting it were mainly marketing ones. The huge carpet manufacturing business in the United States was baulking at the use of wool because of the unevenness of quality, with defects from a number of causes, including second cuts by shearers. Even more disconcerting were price fluctuations caused by the auction system which made planning difficult for carpet-makers and inclined them towards the synthetic fibres for which forward orders at fixed prices could be made for delivery on arranged dates.

Also, New Zealand was floundering for wool marketing skills it had never seriously needed before. It led to the great wool acquisition debate of the early 1970s, with the grass-roots farmer, so to speak, resisting the attempt of the Featherston Street (political) farmers to buy in the national wool clip and control distribution and marketing in a way that had proved so successful in the dairy industry.

A consummate showman, Godfrey toured the world, became an international star and attracted enormous attention with his performances at the World Expo in Osaka, Japan, in 1970. A year later, he and a partner launched the Agrodome at Ngongotahā, near Rotorua, a 160-hectare working farm, featuring farm tasks, including 1950s-style shearing, and performing farm animals, a tourist attraction that long survived him. During his career he made demonstration tours to Britain, the United States, South America and France, was made a 'Hero of the Soviet Union' after a shearing performance in front of Nikita Khrushchev, and Britain's *Guardian* newspaper wrote of him: Godfrey Bowen's arms flow with the grace of a Nureyev shaping up to an arabesque, or a Barbirolli bringing in the cellos…' and continued with more superlatives.

Many skilled shearers had gone before the Bowens since cutting the fleece from sheep rapidly and efficiently had first become of economic importance to New Zealand in the 1840s. The first big-name champion was Charles Smith from Southern Hawke's Bay who completed 200 sheep in a day with blade shears.

Two Māori shearers, Johnny Hape and G.L. Stuart, both also from Hawke's Bay, finessed what had become known as a 'downward belly' technique in the 1920s. A number of other shearers improved their methods as the job became more and more competitive and a Māori shearer from the East Coast took the nine-hour-day record to 433 in 1934, a record that stood for two decades until the Bowens arrived on the scene.

The earliest runholders who pioneered large tracts of country in the South Island had a special range of problems as well as mustering and shearing — wild dogs attacking stock; rats, mosquitoes, fleas, sandflies and blowflies scourging the men and their supplies; the poisonous tutu shrub (mainly near the coast), speargrass and matagouri plants; floods, swamps, bridgeless rivers, pathless hills and mountains; and, always, isolation.

The loneliness was often long and intensive, as Herries Beattie tells in *Early Runholding in Otago*:

> *Most accounts of pioneer life on the runs stress the loneliness. Women had a very lonely time of it. The first woman on the Waikaka Run went there in December, 1858, and lived in a 'warrie' [whare] at what was later the junction of three populous roads, and yet she was there twelve months before she saw another woman.*
>
> *The same experience befell another woman a dozen years later on an out-of-the-way run in the Hokonuis. Two travellers through Otago Central in 1859 saw only one woman. When she saw them in the distance she tore down on horseback to see if, by any chance, her husband was one of them. He had ridden away to get cattle four months earlier, and she had heard nothing of him. When asked if she was not afraid to live alone in such an outlandish locality, she nonchalantly replied, 'Oh, no! There's no one to hurt me!'*

Scab was the first scourge of the sheepman. It came before the rabbit menace and proved easier to eradicate. It was imported originally with Merinos from Australia, was known in Nelson in 1845, the Wairarapa in 1846 and at the Deans' farm, Riccarton in 1848. It spread swiftly through Canterbury and Otago where stock on the drove mingled with sheep on unfenced runs. Infection with the scab mite caused dermatitis that set the sheep to rubbing itself against anything it could find, thus helping to spread the contagious disease. The sheep would rub its wool off and gradually lose condition.

Runholders took to the problem with tobacco and sulphur dips and by inducing the government to invoke stringent laws against those who did not attack the disease or those who spread it by moving infected animals. Scab was at its peak in the 1860s and gave a fresh fillip to fencing, as galvanised wire became plentiful. The tough legislation worked. By 1892 the disease was officially declared eradicated.

An intransigent attitude to a fatal catarrhal disease of sheep, which was rife in Australia at the same time, kept it out of New Zealand and focused the attention of farmers in this country on to the effectiveness of quarantine laws. As soon as the threat of a disease became known, farmers and the government moved swiftly to keep it out. A parliamentary select committee in 1861 met to take evidence on whether the importation of cattle should be prohibited from those countries where pleuro-pneumonia was known to exist. It recommended accordingly. Parliament, by the way, not only acted with alacrity to a threat in those days but dealt with crises expeditiously. The select committee met on 13 June 1861, heard evidence from a number of farmers, butchers and veterinarians, and reported to Parliament on schedule — eight days later.

The government took swift action again in 1866 to prohibit animal imports of any sort from countries known to have rinderpest, an infectious viral disease, at that time raging in Panama. The legislation set up the framework within which emergency measures could also be taken to stop the spread of the disease within the country if it arrived. Thus early advantage was taken of New Zealand's isolation in the world to protect its herds and flocks from the fatal or debilitating effects of many of the worst animal diseases. Successive governments — aware that the economic stability of the country depended on the ready acceptance overseas of disease-free animals and animal products exported from this country — continued an implacable quarantine regime.

By 1858 the flat land in Canterbury Province had been taken up. It was still understocked, but it was accounted for. Settlement by squatting was spreading swiftly westwards to the foothills of the Southern Alps and about the same time was fanning out from Dunedin into Otago and Southland where conditions were often even more rigorous and isolated than in Canterbury. About 140,000 acres of New Zealand (57,000 hectares) mostly in the South Island, had been ploughed or burnt off and sown in pasture in 1858. Twenty-five years later, more than five million acres (about two million hectares) had been sown in pasture and average fleece weights had doubled over the same period.

The big squatters, leasing thousands of acres each, developed a kind of divine-right-to-land approach to politics in Canterbury. When Grey dropped the price of land to five shillings or ten shillings an acre (0.404 of a hectare), depending on locality, the idea was probably to give more men access to land, thus preventing too large aggregations. But the shrewd and sophisticated runholders fought back with spotting (buying the best spots on their huge leasehold tracts to make intrusion difficult) and gridironing (buying strips to make access difficult or in such a way as to leave less than the minimum twenty acres, about eight hectares, for an outsider to buy).

Because they were accumulating capital faster than any other group anyway, the lower prices meant they could amass larger areas than ever. 'It is a strangely ironical fact,' wrote Randal Burdon in *High Country*, 'that at the very time when he [Grey] was making stirring speeches about the evils of land monopoly, the greatest amount of land accumulation was going on.'

As William Pember Reeves put it: 'In the earlier years the squatters were merry monarchs, reigning as supreme in the Provincial Councils as in the jockey clubs. They made very wide and excessively severe laws to safeguard their stock from infection, and other laws, by no means so wise, to safeguard their runs from selection, laws which undoubtedly hampered agricultural progress.'

The intelligence, wealth and arrogance of these men made land the prime political issue right through into the twentieth century and created the powerful farming lobby, dominant nationally until the 1960s. 'The second generation of large graziers rarely inherited either the charm or the education of their fathers,' wrote Alley and Hall, 'but at least they produced an increasing quantity of wool.'

CHAPTER SIX

# Apart from the Good Earth

The settlement of the colony by Europeans, almost entirely of British extraction, was not a seamless move into farming the good earth. The discovery of gold in Otago turned Dunedin into something of an urban town, with secondary industries servicing the needs of the miners. Under the guidance of hard-working Scottish immigrants, it became a prosperous and well-managed city with the highest steady resident population in the country until the end of the nineteenth century.

And while gold and wool dominated the economies of the south, Auckland became an entrepôt and was never focused on Home in the way the farming communities were. The wooded slopes that rose from the southern side of the Waitematā Harbour were decreed to be a new capital city by Governor Hobson in 1840; so it was immediately a government centre and within a short time it became a major trading port. After twelve months it had a population of 1500, not including the province's 59,000 Māori. The energy and initiative of surrounding Māori fed the population and whatever agriculture spread around the city by immigrants was aimed mainly at supplementing this.

The southern settlements were angry about Auckland becoming the capital and they gave Hobson a hard time. They were kitset colonies whereas Auckland was knocked up in a hurry like the raupō whare most of its people lived in. In 1845, the *Nelson Examiner* ranted in the full-throated prose of the journalism of the time:

> *We have asked before and we ask again, how it comes that the government does not publish statistical returns of its own settlements?*

> *Where are we to find an account of the cattle, sheep and crops of Auckland and its dependencies?*
>
> *The last census in Cook Strait was taken by the government; was none taken in the north? If not, why not? If any was taken, why not published? Were they afraid of showing the nakedness of the land — of giving a beggarly account of empty boxes — of proving by comparison of figures the propriety of Hobson's choice and Fitzroy's adoption?*

Some figures on the business of the ports, published by Auckland's *New Zealander* newspaper, showed that in terms of ships and tonnage of exports Auckland was away ahead of Nelson almost from the first day of its existence and that by 1844 it had passed Wellington in numbers of vessels, but not in export tonnage.

Auckland was not sheep country and milk, butter and cheese production was only as big as it needed to be to service the local people. It was even then a commercial town, a thriving port. In that sense it also had the characteristics of an urban rather than a rural settlement.

It was surrounded by forest, had no Wairarapa at its back door nor any huge tracts of tussock-covered plains, so sheep on any scale were impracticable. Agriculture pushed back through what is now Epsom to the south, and crops were grown, cows milked and steers slaughtered to feed the locals, to meet the demands of new arrivals, including British troops brought in to fight against Māori, and to supply the ships that came from around the Pacific.

The city had closer ties with Sydney than with other centres in New Zealand. In 1848, for example, 1430 people arrived in Auckland, more than half of them from Australia (699 from New South Wales and 632 from Britain). Three hundred and seventy-two left the city (306 of them for New South Wales and twenty-five only for Britain).

Auckland didn't have the urgent need for wool to export because it had kauri gum and timber. These extractive industries certainly held back its agriculture. An Agricultural and Horticultural Society was formed in 1843, but it hardly thrived. Some years no show was held. One of the men who regularly collected a prize for sheep in the 1850s and early 1860s was John Grigg, the man who later moved to

Canterbury and became a legendary sheep and wheat farmer. For aspiring pastoralists, south was the way to go.

By the 1880s, the Auckland Agricultural and Horticultural Society had become, grandly, the New Zealand Agricultural Society and then, humbly, the Auckland Agricultural and Pastoral Association which was in such serious financial trouble that no show was held for three years until a burly Māngere farmer, Bill Massey, became president and reorganised it — a light administrative workout for a later, long-time prime minister.

Through the 1850s, returns showed there were generally as many cattle as sheep in the Auckland region. Flocks were small and mostly aimed at the domestic meat market. Wool was only occasionally listed among exports.

In tone and character, Auckland was a metropolitan Pacific port. But in 1850 a perceptive correspondent wrote a prophetic article for the *Maori Messenger*:

> *At an early period of the publication of this journal we compiled a mass of intelligence on the subject of dairy farming and endeavoured to show to what great and beneficial results it would lead if the native landed proprietors could only be brought to turn their attention to a branch of rural industry which, whilst it became the means of reclaiming and enhancing the value of their waste lands, would gradually pave the way for a great and lucrative source of immediate prosperity.*
>
> *The Northern Island of New Zealand is peculiarly adapted for dairy farming. Its pastures are rich and nutritious in a remarkable degree. There is ample sufficiency of moisture throughout the year to sustain and nourish vegetation, and we need but point attention to a well known fact, namely, that the Tamaki meadows have throughout the year maintained a hundred oxen on a hundred acres of ground to prove the capacity of the country, and to show that New Zealand might profitably become the dairy farm and market garden of the Australians. ... The small supplies hitherto shipped from here have realised excellent prices and have commanded the highest character. Here, then, is a mine of wealth in which our native yeomanry would do well to dig.*

By 1852, exports of butter and cheese from the port of Auckland earned £456, half as much as kauri gum but only about one-third as much as potatoes.

Auckland grew as fast as it needed to. Land development was expensive. It wasn't until the next century that the full agricultural potential of the northern half of the North Island was fully understood.

In the south it was different. The offspring of many of the graziers who had special talents or academic skills defected to England, which they still thought of as Home. This sense of identity with the British lasted long in New Zealand. The poor and rejected who formed substantial numbers of the immigrants to Australia and the Americas had less cause to remember their origins than the significantly better off, more respectable 'colonists' of New Zealand. It was part of the national state of mind and consequently part also of the national literature, as an early poem by Mary Colborne-Veel illustrates:

*Homely flowers set*
*Where our farmsteads rise,*
*Make an England yet*
*Under sunny southern skies.*
*Lilac scent is blown*
*With wattle on the breeze;*
*September bids the leaves grow broad*
*On happy English trees;*
*And apple-orchards smile again*
*In sweet, familiar show —*
*But in my heart is mourning*
*For the scenes of long ago.*

By centennial year, 1940, most New Zealanders had been born here, but so many church, political and academic leaders were from Britain that the British thing was still pervasive. Writers such as Robin Hyde and A.R.D. Fairburn were indelibly, if a little self-consciously, New Zealanders first and their literary voices reflected that; but they belonged to a kind of subculture. Auckland University College published a book in 1940 called *1840 and After*, and subtitled *Essays Written on the Occasion of the New Zealand Centenary.*

It may seem extraordinary now, but most of the authors, all from the staff of Auckland University, ignored this country's existence. In his foreword, the college president, Mr W.H. Cocker, said: 'English culture of the early and middle nineteenth century has a special significance for New Zealand and her history. The authors of this book … have therefore turned their telescopes upon Victorian England. What kind of civilisation was it that was brought to this country? … The reader of this book will have a reasonably complete picture of what was probably the most significant period in New Zealand's past.'

Earlier, in Canterbury, but also later in Hawke's Bay, runholders who achieved success and affluence had tried to emulate the kind of life lived by the landed gentry of England. They achieved a sort of elegance, the more impressive for its remoteness from civilisation. As Lady Barker put it in *Station Life in New Zealand:*

> *The older I grow the more convinced I am that contrast is everything in this world; and nothing I can write can give you any idea of the delightful change from the bleak country we had been slowly travelling through in pouring rain, to the warmth and brightness of this charming house [near Amberley]. …*
>
> *The inside of the house is as charming as the outside, and the perfection of comfort; but I am perpetually wondering how all the furniture — especially the fragile part of it — got here.*

She describes a picnic that is a sedulous transference to New Zealand of an English manor house ritual. But elsewhere she refers with kindly condescension to the poor quality of servants in Christchurch, the inadequate music and unsophisticated behaviour at balls.

Many runholders aspired to a high life because of their background, but they had paid their dues, as Alley and Hall said:

> *The men who took up the early sheep stations were often people of education, belonging to families of English country gentry, or at least to those connected with the services and the professions. In the early years these men did not hesitate to put up with extremely crude living conditions — the slab or cob hut thirty miles from a neighbour. Men reading Latin or Greek for pleasure lived the hard, self-sufficient life of the back country with gusto, still hoping great things for the future.*

A kind of egalitarianism nevertheless developed, not as a philosophy but as a fact of the existence of the new colony, a fact commented on by many early visitors. Often the motive for emigration, even among the middle and upper middle classes, was a social incompatibility at home in Britain, some rejection or failure, and no matter how much the actual decision to come out was made by the emigrant personally, it was still a kind of exile.

A figure from the folklore of New Zealand was the stereotype mystery man of good education, working as a station hand, provoking speculation that he was of noble descent, or at least from the upper class, because he spoke so well. He would go to town periodically and drink away his earnings, or his remittance from Home, then come back to the station and work hard until next time.

The Remittance Men were members of wealthy British families who were sent to the most distant of colonies because they had committed a crime or were generally troublesome and a threat to the family reputation, mostly because of alcoholism. They would periodically line up at the post offices and banks to receive money remitted from families back Home.

These, and others among the swaggers, were the Remittance Men, members of wealthy British families who were sent to the most distant of colonies because they had committed a crime or were generally troublesome and a threat to the family reputation, mostly because of alcoholism. They would periodically line up at the post offices and banks to receive money remitted from families back Home.

'Wonderful tales are told of cultivated men in the wilderness,' wrote William Pember Reeves, 'Oxonians disguised as station-cooks, who quoted Virgil over their dish-washing or asked your opinion on a tough passage of Thucydides whilst baking a batch of bread. Most working settlers, as a matter of fact, did well enough if they kept up a running acquaintance with English literature; and station-cooks, as a race, were ever greater at grog than at Greek.'

When transport and communication were slow and inevitably imprecise and arduous because of the intervention of so much time, it was easy for people to dramatise themselves if only by innuendo, or by what they didn't say. Facts were dilatory travellers to the advantage of romanticisers, dreamers, frauds and swindlers. According to *The Age of the Great Sheep Runs*, an essay by P.R. Stephens: 'There was little that was patriarchal in the relationship of the runholder to his employees and on many properties a somewhat suspicious, harsh attitude, probably imported from Australia, was the rule.'

Accommodation was poor, and the food often little more than unlimited supplies of tea, bread and mutton; for the whole system was based on the existence of a large number of itinerant casual workers. Shearers were in a better position to assert their rights and some fairly bitter disputes took place. The Amuri Sheep Farmers' Association during the late 1880s even asked the government to import shearers from Africa.

But a code of hospitality often did prevail with food and rough shelter made available to itinerants by some farmers and their families. The life on the road as a casual labourer or shearer was spartan, though — men mostly without women, and little social life outside the pub, and with alcohol used freely to numb despair and blunt aspirations. Alcohol became an enormous problem affecting the health and well-being of thousands. Accounts of early life in New Zealand often refer to alcohol as the major health problem of the time.

The gravity of the problem led to the rise from the earliest days of European settlement of a vigorous Prohibitionist campaign intent on removing alcohol from society. It

The life on the road as a casual labourer or shearer was spartan, though — men mostly without women, and little social life outside the pub, and with alcohol used freely to numb despair and blunt aspirations. Accommodation was poor, and the food often little more than unlimited supplies of tea, bread and mutton.

was this politically charged movement that contributed throughout the nineteenth century towards granting of votes for women. Prohibitionists considered that female enfranchisement would lead to their victory because women were most commonly the victims of male drunkenness. And they almost succeeded in 1919 when only the vote of soldiers turned the vote against Prohibition.

For many decades from the last quarter of the nineteenth century, itinerant farm workers followed the seasons in Australia and New Zealand. They were swaggers — on the swag. The swag contained a small tent, a sheet of oilcloth to cover the damp ground, a billy for tea, and all the rest of their worldly possessions.

New Zealand politician and writer John A. Lee was on the swag as a young man and wrote of the life in a rich and funny book, *Shining with the Shiner*:

> *They came regularly from Australia for the shearing, the harvesting, in large numbers. ... They walked to work. It was a cheap way to get by. If there was no work and no dole, time on the road did not much matter. A fortnight would take a man across most of the South Island.*
>
> *They erected calico tents in the trees near the fresh water and were ready to start when the season and the boss willed. ... And the concertinas might come out around the fire and voices might sing 'Alice Ben Bolt' or 'The Wild Colonial Boy'.*

Lee was one of many storytellers of the time, encouraged as they were by sitting around the fire in the evenings with no entertainment but each other. It was a hard life between drinks. Runholders hired special workers to set the pace and those who couldn't match it were soon sacked. Lee wrote:

> *And as the country grew settled and the telephone and good communications came and as the steady among the industrious, for all were industrious if some were improvident, settled into jobs ... The men who remained on the road were the men who drank as hard as they worked, the tough and strong who were the cageless, the unadaptable, the artists who in a strange country had become labourers; and of this type there were many...*
>
> *In those days the gap between vagabondage and respectability*

*was not immense, a bad season, a foreclosed mortgage. How many a prominent citizen of yesterday have I see settle down, fill a pipe, and start: 'When I carried the swag...'.*

Lee, on the road as a boy after escaping from borstal, was the most famous and the most published of the New Zealand itinerants and, like many of them, got off the road to find steadier work. The experience of living rough with hard-core alcoholics encouraged him to be a teetotaller, reinforcing an anathema to drink built up during a troubled childhood in Dunedin.

An even larger tradition of the itinerant storytelling swagger spread through Australia. Many men travelled among the states and included New Zealand in days when their common remoteness held the two countries closer together than today.

One of the last of the swaggers was 'Russian Jack', still on the road in the Wairarapa until the beginning of the 1960s, by when he was what we would today call simply 'homeless'. He died in 1968 aged ninety.

## Chapter Seven

# Grain and Ingenuity

The gold rushes to Central Otago and, later, Westland, and sliding wool prices brought about a move to wheat growing and a boom in other grain crops as well. The Free Church of Scotland settlers of Dunedin had been wary of immigration in the 1850s from what they saw as undesirable sources, and those sources included Catholic Irish and anyone from the waning goldfields of California and Victoria.

But Central Otago had already attracted prospectors because the big buckled landscape with fast-flowing rivers looked promising to the practised eye. After Tasmanian Gabriel Read found substantial deposits of alluvial gold at Tuapeka in 1861, the quick prosperity it brought to a province in a slump soon persuaded the people of Otago gold was no bad thing at all whether it attracted ruffians or not.

The first report of gold in the province arrived in Dunedin in 1851 when the settlement was only three years old. Two men looking for grazing land claimed to have found some. The first firm discovery was in 1856 when the Surveyor-General, Charles Ligar, said he had found widespread traces in sand and gravel during a visit to Mataura. Glimpses of 'colour' were also seen in other parts of the province during the rest of the 1850s. There were no big finds, though. The land was coquettish.

Dunedin's settlers heard Ligar's news with what Otago historian Alexander McLintock called 'ill-concealed alarm'. The Scots wanted to preserve their settlement from temptations within and shore it up against corruption of their values from without. When gold was not more than glittering talk, settlement leader William Cargill and his sober, upright cohorts could foresee the boots of ten thousand diggers trampling their cultural and religious purity. Yet when Read made his discovery, half the sober population of Dunedin was off and running for Tuapeka.

Gold had an enormous impact on the farming of New Zealand. The population of Otago in 1861 was a shade over 12,000. Ten years later it was close to 60,000. More than £23 million of gold had been pulled out of the ground and the hinterland of Otago was opened up — in some cases left scarred beyond recognition. The influx of labourers and artisans into Dunedin liberated social and political thinking from the dominance of the conservative churchmen and the pastoralists whose power had grown through the 1850s. Gold also consolidated the economic superiority of the South Island over the North, already established by the sheepmen. It wasn't until the turn of the twentieth century that the North Island began to catch up.

The population of the whole country during the golden decade moved from around 60,000 to 250,000. With new national wealth from the export of gold (which peaked at £2,845,000 in 1866, more than wool earned that year) and tens of thousands more people, farmers could shift a little from growing wool for export to producing for the local market. Around the goldfields, agriculture had swiftly followed the population build-up with a sudden stretch of acreage under consumer crops.

It wasn't just for food. Barley production soared as men without women — the miners who crowded in — found solace in booze. The ratio of women on the Central Otago goldfields was one to one hundred men in 1861, although it grew to 18:100 in 1864, 30:100 in 1867 and 47:100 in 1871. (Over the whole of New Zealand in 1861 women numbered only sixty to a hundred men.) Seventeen breweries were operating in the province by 1871, producing more than forty per cent of the country's output of beer.

So gold and its direct effects, along with problems managing sheep, price fluctuations for wool, and the development of cultivating and harvesting machinery meant grain waved across the South Island countryside in the 1870s, 1880s and 1890s. South Canterbury farmer and poet F.H. Woods wrote:

*The seas of wheat o'er which the wind-made wavelets*
*Sweep on and break upon a phantom shore.*

The first cereal crops in New Zealand were harvested with sickles and scythes. The first mechanical mowers cut swathes along the paddocks in the 1870s, but the sheaving was still done by hand for another decade when reapers and binders made their appearance and steadily developed in capacity and speed. The equipment encouraged arable farmers to cultivate more carefully to reduce the number of weeds harvested by the unselective machines. Tractors gradually took over the task of pulling the machines, but horses still commonly did the job until well into the first third of the twentieth century. Header-harvesters appeared not long before the Second World War.

In the nineteenth century, Canterbury and Otago were the provinces — still are for that matter — most suited to the growing of grains. Farmers there had the climate, the soil, the flat acreage, the capital, the organising ability needed for large-scale cultivation, and access to the extra labour needed for wheat. And they had the big thinkers, the risk-takers. One of the virtues of early New Zealand farmers was that they had no traditions to tie up their minds. Many of the pioneers had little previous experience and what they had needed testing against a new environment in a new country.

Otago's Donald Reid, who left his lowly home in Perthshire at the age of fifteen, had twenty prime acres at Forbury, in Dunedin, at eighteen, 200 acres at twenty and within twelve years of his arrival was cultivating 500 acres. With farm produce, general stores and fast transport he pulled more money out of the gold rush than all but a few of the diggers — and he even got a share of the actual gold. As his farm labourers prepared to hot foot it for the goldfields, Reid counselled patience, formed a co-operative group and they all got their share from a profitable claim worked on their behalf.

Characteristically, Reid was rapt in the potential of machinery for cultivating and harvesting grain: 'In 1861, he [Reid] bought a reaping machine,' wrote Alley and Hall.

> *It belonged to that strenuous old type from which the cut grain had to be raked off sideways. He soon replaced this with a back-delivery reaper.*

*In 1862 he used his profits from gold, including a share in an actual claim, to buy machinery in quantity — a chaff-cutter, corn-crusher, a threshing machine and a steam engine. A more powerful steam threshing plant was already at work in the Taieri, but Reid preferred to be able to do his own work in his own time and to hire out his machinery to others.*

From about 1865 he gradually changed over from cropping to sheep farming, his first highly profitable herd of cattle having suffered a good deal from stock disease.

This flexibility is typical of the most alert farming minds in early New Zealand. Reid and others resumed the production of grain on a large scale in the 1870s, when it developed into a substantial export industry. Meanwhile the pastures he laid down in English grasses carried increasing flocks of sheep.

John Grigg and Duncan Cameron were two of many Canterbury men who backed their practical skills and foresight by moving into wheat when the time was right, and back into sheep again when refrigeration looked a prospect. Not that sheep farming suffered greatly during the wheat boom. Flock numbers kept up and fleece weights for each animal doubled during the last quarter of the nineteenth century. The families and estates of these agricultural entrepreneurs may now be institutions firmly grafted to the establishment, but in their day they were the highest fliers New Zealand has produced in business. Their imagination and courage were applied to big risks and big stakes. They won most of the time and New Zealand shared the prize. Rich and resourceful John Grigg was always at the forefront of mechanical and business innovation in Canterbury and his large wool clip went to market in style, powered along a dirt road by a huge steam engine.

Like the Americans of their time, these Canterbury and Otago farmers had great faith in machinery, and they had the support of imaginative blacksmiths, mechanics and engineers. Interest was so intense, a reaping match was held in Hagley Park, Christchurch, in 1864, to decide on the efficiency of six different types of imported and home-made machines. Each had to cut one and a half acres of a fairly even crop of oats.

First prize went to a Burgess and Key (imported) hand delivery reaper which beat the first McCormick reaper. (They were reapers and not reapers and binders.) The first reaper and binder was imported in 1877 by a Christchurch blacksmith called John Anderson and, according to an article in the Centennial Supplement of the Christchurch *Press* in 1940: 'These machines did away with labour and tales are told of barns being broken into and machines being broken up by labourers who feared they would soon be out of work.' But resistance from the Luddites was slight. In the same year, the first American seed drill arrived. Business was brisk. The records of one firm (Morrow, Bassett and Co.) show that fifty McCormick harvesters were brought out in 1879 and 1880 and, even though the price was high at £75, the supply was not enough.

Anderson set up as a blacksmith in 1850, and in 1857 began making farm implements, many of them to his own design. He sent his two eldest sons to Edinburgh in 1866 and they returned a few years later with degrees in civil and mechanical engineering. He was a man of great diligence and high imagination: during his blacksmith apprenticeship days in Edinburgh, he worked from 6 a.m. to 6 p.m., but found time to gain a diploma from the city's School of Arts.

Wool prices were wavering at the beginning of the 1870s; and during the 1860s farmers, especially in Canterbury, had their first serious climatic difficulties getting stock through the year in open grazing. After the benign fifties, two sheep-burying snows and a killer flood made the decade arduous and ruined some runholders. So ploughshares bit deep into the better land at the beginning of the seventies and the 'bonanza' wheat farming began, almost ninety per cent of it in the

The first reaper and binder was imported in 1877 and, according to an article in the *Christchurch Press* in 1940: 'These machines did away with labour and tales are told of barns being broken into and machines being broken up by labourers who feared they would soon be out of work.'

South Island. Not that the price was encouraging; it was actually down as the 1880s began and as it went down the acreage went up.

The 1880s decade was a time of economic depression. In this first period of wheat's 'high importance', wrote F.W. Hilgendorf in *Wheat in New Zealand*, prices averaged only 3s 4d (34c) a bushel, over a twenty-year period, 'yet wheat growing paid better than anything else. Cultivation had become mechanised with the introduction of the drill and the binder, and thus large wheat growing farms came into existence.' John Grigg's Longbeach property had 3000 acres under grain, 120 working horses, and twenty ploughs of two or three furrows, working on one field. Duncan Cameron of Springfield Estate, near Methven, grew wheat on a comparable scale, and many others approached it.

Longbeach was a property that attracted huge attention. John Grigg was New Zealand's most famous nineteenth-century farmer. His farm was so much his alter ego he was referred to almost exclusively in one breath as 'John Grigg of Longbeach'.

A family narrative was that the Griggs were descended from two Scottish brothers who migrated to Cornwall when the McGregor clan was dispossessed of its land and civil rights in the seventeenth century. John Grigg was born on the family's Brodbrane farm, near Duloe, in Cornwall in 1825. He could have stayed on the family property and lived something close to a manorial life but chose to sell up and sail for the colonies. His decision was prompted by a desire for adventure garnished by the departure for New Zealand of Martha Vercoe, with whom he had fallen in love.

Grigg left England in 1854, stayed for a few months in Melbourne, sailed for Auckland, leased a farm at Ōtāhuhu, and in June 1855 married Martha Vercoe. In Auckland, he successfully grew potatoes for export (after one early failure) and profited from hay sales for army horses as British soldiers arrived for the wars in Taranaki and the Waikato. He imported stud sheep and by the time he decided to move to Canterbury in 1865, he had built up a flock of purebred Leicesters that drew attention from throughout the colony when it was dispersed at auction. He was a lay reader and one of the signatories of the original constitution of the Church of England in New Zealand.

He decided to acquire a property already known as Longbeach. The Hinds River ran from hills to the west across shingle beds and then disappeared into a large swamp before it reached the coast. On the northern side of Longbeach was the Ashburton River. Surveyors called it an 'impenetrable swamp'. Graziers ran a few cattle over the drier parts but otherwise kept clear of it.

John Grigg freeholded 2000 acres in 1864 and, by 'gridironing', bought 32,000 acres over the following seven years. Gridironing, as described earlier, was a process by which a settler bought strategic blocks within a large area to render small blocks uneconomic for anyone else to retain or buy. He moved south in 1866, bought a house for the family in Christchurch and then lived in a sod hut as he audaciously dug a bed for the Hinds to run into the sea, thus beginning to drain the swamp. It was a remarkable feat that took a long time, but he once said, 'On Sundays I used to go and lie down on the beach listening to the sound of the waves and dreaming of what Longbeach should one day be.'

Grigg stopped buying in cattle for fattening, established a Shorthorn beef cattle herd, founded a Devon cattle stud and then began droving cattle through the Rakaia Gorge to Westland where butchers in the goldmining towns paid a good price for them. But the trade waned as did the whole beef cattle market by the early 1880s and Grigg, with foresight for which he became famous, had already changed to sheep ready for the export trade in mutton that grew from the introduction of refrigerated shipping in 1882. With a conviction that dairying would become a major industry, he imported Holstein cattle from England and expanded a prime milking herd.

By 1880, the farm was stocked with 10,000 Merino and half-bred ewes using Leicesters and Lincolns and even the special mutton breed, Shropshires, as he set out to breed sheep to serve the British meat market before refrigerated shipping techniques had been proven.

Grigg was an enthusiast for purebred stock. He had Clydesdale and Thoroughbred studs among 400 horses on the property and sold about eighty horses a year. The farm also carried more than 2000 pigs. Meanwhile, Longbeach had a cropping programme which had begun the 1960s. By 1879, 3000 acres were under wheat, oats and barley; 1400 under roots and forage; 500 under grass for hay; and 300 under peas.

Seventeen years after Grigg first set foot on the property, a village surrounded the homestead with 150 full-time staff, including shepherds, dairymen, teamsters and other stockmen, fencers, gardeners, general farmhands, as well as bakers, blacksmiths, carpenters, saddlers, butchers and bacon-curers, and those involved in boiling down surplus sheep carcases into tallow.

He convened the meeting in November 1881 that formed the Canterbury Frozen Meat Company and became its first chairman of directors, a post he held for nearly twenty years. In one season during the 1890s, he bought 80,000 sheep and lambs for fattening and exporting to London. In one day, he drafted 4400 lambs on Longbeach for the freezing works and the average carcase weight was over forty-two pounds. In 1894, 4500 acres were under wheat.

The leader of this extraordinary community, paying an estimated £8000 a year in wages, was a fair-haired man of average height, limping slightly on a leg damaged in a childhood accident, his face covered by a full beard, always on the move on foot or horseback. Grigg's organising and administrative talents were almost as legendary as the foresight that enabled him so often to predict trends and market developments. His reputation was for unalloyed honesty and remarkable generosity to his loyal staff.

His obituary in *The Press*, Christchurch, in 1901, said Longbeach 'has been described by a widely-travelled expert as the best farm in the world. It has long been one of the show places of the colony, regarded by the people of Canterbury with pardonable pride...' In less than forty years, 'he converted a tract of country, some thirty thousand acres or more in extent, which even hardy surveyors termed "impenetrable bog" into broad paddocks which have produced enormous crops of cereals and on which have fattened multitudinous flocks and herds. ... He lived his life usefully and strenuously, and now has passed away, full of years and honour, leaving the colony at large richer for his labours and his fine example.'

Despite mechanisation, wheat growing was still a labour-intensive business, even compared with wool growing which had its annual peaks of mustering, dipping and shearing. Shearing required additional, hard-working labour for a short period and it was the major organisational problem that a runholder built his year around. 'When this great business of the year is over,' wrote Alexander Bathgate in his *Colonial Experiences*, 'the squatter often allows himself a holiday, and after shearing makes a run to town. His calendar appears to circle round this epoch, and events are spoken of as having occurred so long before or after shearing.' In normal times, he noted, 'the station employees are few in number and vary according to the size and ideas of the owner. Two or three shepherds, invariably Scotch, and an old man or two, form the usual staff.'

That was written in the early 1870s. By the eighties, recession had set in and a pool of labour was available for the more demanding task of grain growing. And this labour was seriously needed because harvesting, stacking and threshing wheat, then bagging and storing it, required even greater labour resources, especially if the farmer was fighting time, against rain or its likelihood.

Wheat had been grown for many years in New Zealand, at first in small fields for tribal or missionary settlement subsistence, but some was exported to Australia from the earliest times. The area under the crop in 1861 was only 30,000 acres and falling. Gold had something to do with that at first, attracting labour away from the farms. But during the second half of the decade, stimulated by gold-rush demand, the acreage began to climb. It increased between 1870 and 1880 from 86,000 acres to 157,000, and during the early 1880s to 321,000. By 1881, forty small flour mills were in operation in Canterbury, but the number diminished as large steam-powered mills took over.

The acreage was high again in the first two decades of the twentieth century and climbed again to 160,000 hectares in 1920–21. Most years since then, New Zealand has had to import up to one-third of its wheat consumption and growers have had to match the high milling quality of the Australian product. Despite a serious price slump in the 1980s, a number of South Island (particularly Canterbury) growers persisted with wheat, supplemented with other grains and with grasses for seed as well as sheep. The area sown in recent years has been around 70–90,000 hectares and experienced growers with high-quality varieties have yields twice as high as before the Second World War.

Grain production peaked in Otago in 1890, when 376,000 acres of oats were harvested as well as 294 acres of wheat. But as with wheat, yields did not hold up well until better prepared seeds of improved strains were reared under more skilled management.

New Zealand was growing nearly an acre of wheat for every person in the country during the 1880s and 1890s and net exports reached 5,142,000 bushels in 1883–84. Yet the yield was low and getting lower. 'Wheat growing either spread into land that was not suited for it,' wrote Hilgendorf, 'or else the same land must have been used for wheat growing either continuously or in a very close rotation'.

Part of the land now occupied by Lincoln College was cropped with wheat for thirteen years in succession, after being broken in from the tussock, and indeed such a practice is hardly surprising, seeing that the farmer of those days had little other outlet for his activities. The population of the country was still small and the internal demand for such crops as potatoes trifling, but, above all, there was no frozen meat trade, and so no fat lamb industry until the mid to late 1880s. Nothing could be exported except wheat and wool, and so wheat was grown year after year. There were no manures and little rotation of crops, and so the yield, starting low, went down and down.

The population of the country was still small and the internal demand for such crops as potatoes trifling, but, above all, there was no frozen meat trade, and so no fat lamb industry until the mid to late 1880s. Nothing could be exported except wheat and wool, and so wheat was grown year after year.

Yields may have been abysmal at around twenty-three bushels to the acre by 1900 compared with what the land had been capable of, but in global terms they were not low: New Zealand was still in the top half of the world's wheat producers in relation to yield.

Exports of all grains reached 13.3 million bushels in 1901, up 3.5 million from the year before, according to the figures issued in 1904 by the Registrar General Edward von Dadelszen, and then the crop declined spectacularly to less than six million bushels in 1902. In 1908, the Minister for Agriculture, Robert McNab, an historian as well as politician, wrote:

The yields of grain are the highest in the world with the sole exception of those of Great Britain but so successful are sown grasses and forage plants and so excellent is the pasturage of the native grasses that the production of wool, meat and dairy produce has proved more profitable than grain growing and exports consist of (1) wool (2) frozen meat and (3) dairy produce; while the cultivation of wheat and oats has been reduced to the limits of domestic requirements.

The see-sawing of wheat acreage was the result of fluctuating prices, which, of course, reflected demand. World prices were below local returns, so production in excess of local demand had to accept the lower return received for exports.

So the tide turned against grain about the same time as the century did. As the fat lamb trade with Britain grew, much of the marginal cropping land was put down in English grasses and the concept of mixed farming spread. Wheat was rotated with feed crops such as rape and turnips. Many of the large stations had been subdivided and farmers with small holdings worked the land more intensively and more sympathetically. By October 1907, McNab reported, land holdings of one acre or more were 73,367, almost double the number of 1890, which was just before the Liberal Government began breaking up the big estates. Many smallholders on quality land had turned to dairying, although export dairying was still hampered by inconsistent quality because of the inadequately equipped cowsheds and factories and the varying ability of farmers and factory managers.

Wheat competed in a free market with farmers sniffing at any suggestion of control as socialism, until 1914 when a drought virtually destroyed the Australian crop and created a crisis that forced governments to intervene and growers to acquiesce in holding prices down. From then on, until recent years, governments forced a measure of control to ensure constituents had access to the staff of life at reasonable prices. The relationship among growers, manufacturers and governments during these decades was never easy and conflict forced the marketing agencies into many shapes and forms as the market became distorted in one way or another. The end of any regulation came in the 1980s when the radical Labour Government cut farming supports. Some wheat farmers never recovered.

About fifty varieties of wheat were grown in New Zealand from when the first seed was introduced from India, through Australia, in pioneering days. Three varieties dominated the wheat fields of the nation from 1860 to the end of the century — Tuscan, Hunters and Pearl, grown in roughly equal proportions. A new strain of Tuscan then took over more than sixty per cent of the market until coming up to the Second World War when the Wheat Research Division and later the Crop Research Division of the Department of Scientific and Industrial Research began releasing local adaptations. Hilgendorf, Cross 7, Aotea and Arawa became staple varieties, with Aotea winning more than eighty per cent of farmers' preference within two or three years of its introduction in the 1950s. During the 1970s, Aotea was followed by Kopara, Takahe and Rongotea.

Superphosphate began to be used around 1910 and from 1920 tractors became available and better prepared the soil and provided more consistent and faster drilling and harvesting. The period between harvest and seed time, when the ground is often very hard, was short for the amount of work that had to be done with horse teams. In a capricious climate, when rain could seriously damage a crop, fast harvesting as soon as the time and grain are ripe is a huge advantage. Seed treatment for disease control and the development of varieties to suit local soil types and microclimates have all helped pull New Zealand's yield up past fifty-five bushels an acre in the best years, among the world's best after Europe where cropping was even more intensive.

Despite excellent and increasing yields — and techniques such as artificial drying of the harvested grain which have spread the potential land available for wheat — this country's scale of operations cannot match the enormous mechanised production from the vast if comparatively low-yielding areas of Australia, North America and

Russia but, more importantly, in most years there are easier ways for most New Zealand farmers to make a living.

Many farmers in Canterbury would make an economic sacrifice rather than move back into the labour-intensive growing of wheat with its need for more meticulous organisation and the risks of good and bad seasons. A researcher after the Second World War reckoned that the wheat-growing area of Canterbury could be raised to 330,000 acres consistently without soil deterioration, and including other areas round the country he claimed the annual area could be lifted to 400,000 acres, yielding more than twenty million bushels a year. But even if the government-set price was lifted above reasonable expectations for fat lamb returns, it was doubtful that farmers would oblige.

What is true of the history of wheat in New Zealand is broadly true of oats, except that it held up as an important crop past the period of wheat decline at the turn of the century because of its value as stock food, and because it can be grown over more of the country than any other cereal. Oats had a following from porridge-loving Otago Scots, but that market was met by a small amount of production. Even today when cereal fibre is a growing diet choice, the demand can be satisfied with crops from less than twenty thousand hectares, most of it in Otago and Southland. Oats retains value as a supplementary greenfeed for livestock. The highest area sown in oats was in the 1890s through to the beginning of the First World War when it reached around 300,000 hectares. Most of it was used as the staple, high-energy food for the horses which were the main means of transport and were breaking in the land.

Barley also had an erratic history of moving from undersupply to surplus and back again until recent years when yields and prices have led to an expansion of production, some of it for export. It was used mostly for malting for the manufacturing of beer, but crops expanded in the North Island — specially Wellington, Hawke's Bay, Bay of Plenty and Auckland — during the second half of the twentieth century, most of it for use as stock feed, either threshed for pigs and poultry or greenfeed for other livestock.

Worldwide, maize ranks in importance with rice and wheat as a cereal with huge and increasing acreage and yields, particularly in the United States. When it was introduced, it was the carbohydrate plant Māori most valued, after potatoes. They soaked it in water and ate it raw, especially the young cobs, and carried the dry grains as travelling food. Maize has been mainly a North Island crop in this country, produced over many years with steadily increasing yields from about 5000 hectares — about half of it for green forage and half for grain.

An explosion of production took place from the late 1960s in the Hawke's Bay, Bay of Plenty, Poverty Bay and Waikato regions, mainly in response to more productive and profitable varieties which have resulted in extraordinary yields in this country. It helped meet increasing demand for starch and vegetable oils, for poultry and pig feed, and for the fattening of cattle which eat it as it stands.

CHAPTER EIGHT

# Brer Rabbit and Other Undesirables

Outside New Zealand and Australia, no creature has had a better press historically than rabbits, except maybe fairies; although it is acknowledged there are bad fairies, but no bad rabbits. They are soft and cuddly and are reluctant to bite or scratch.

So why do New Zealand farmers hate them? Arnold Wall explains in his 'Song of Brer Rabbit in Central Otago':

*Where the sheep fed, there feed I,*
*Depleted lands behind me lie,*
*Of dogs and guns I take no heed*
*I only breed and breed and breed.*

*At traps and guns and dogs I smile,*
*I laugh at cats' and weasels' guile,*
*I frolic, nibble, frisk and feed,*
*But all the time I breed and breed.*

*They send my skin to clever folks,*
*Who turn me into sealskin cloaks,*
*A sorely hunted life I lead,*
*But still I breed and breed and breed.*

*They chase me round with mongrel packs*
*And bear me home in bulging sacks:*
*I smile, and say my simple creed*
*Which is: 'I breed and breed and breed'.*

*When Nature gave the brutes their law*
*To some she gave the tooth, the claw,*
*The poison-fang, the lightning speed;*
*Me she bade only, 'breed, breed'!*

*When foes attack I cannot bite,*
*I have no spirit for to fight,*
*By other methods I succeed*
*I merely breed and breed and breed.*

*The Nation which would keep her place*
*And beat her rivals in the race*
*This little ditty first should read*
*And then go home and — follow my example.*

In February 1777, Captain James Cook released four rabbits on Motuara Island in Queen Charlotte Sound, Marlborough. The breed and origin of these rabbits and how they fared is unclear, but this first introduction was followed by other instances in pre-colonial times.

A pair of rabbits was among the prizes for a ploughing competition at Waimea in 1843. The records of the 1849 Auckland Agricultural and Horticultural Society mention that among the livestock: 'Dr Courtenay exhibited a remarkably fine pair of rabbits.' It was part of the sentimentality of the British immigrants trying to transplant familiar plants and animals, the same sentimentality that brought in other exotic mammals which turned out to be costly pests and that brought in gorse and other noxious weeds.

The importation of rabbits was ill-considered because their depredations in other lands, notably Australia, were already known by then; but it was understandable because the ordinary Englishman and Scot knew the rabbit only as a useful, controllable animal in their homeland. Many of them brought out other plants and animals to this country's enormous benefit.

Gorse became a scourge as it spread like a yellow flame through the country, claiming land from grass, but at least until the turn of the century it had its supporters — and official supporters at that. The Department of Agriculture's officer in charge of the Auckland Province in 1897 said in his annual report:

*Gorse-growing for feeding sheep is being considerably extended in the Bay of Islands. The energetic owner who initiated the growing of gorse for fodder is gradually overcoming prejudice, and the next step will be others following his methods.*

*This is a most interesting development in utilising country that until now has been comparatively wasted, and if an experimental station is established here one of its most interesting features should be the number of sheep per acre grazing in its gorse paddock.*

The first few rabbits brought in either didn't survive or didn't do well enough to spread beyond local colonies. Rabbits were undoubtedly around this country for more than twenty years when, writer and historian Randal Burdon reports, 'one of the passengers in an immigrant ship that arrived at the Bluff in 1864 ... brought out with him some rabbits as a speculation. He sold them at high prices and they were turned out in the vicinity of Invercargill.' They dug in first on the sandhills near the town, but it appears they took some time to adapt to the nearby heavy country. Once they hit the hills though, and the drier country in the middle of the province with its light sweet native pastures, they found their Eden.

Something of a lemming-like inexorability drove them northwards accompanied by a locust-like way of eating out everything as they went — everything a sheep would want to eat, anyway. They grazed pasture right down to the ground, closer than any other grazing animal, and the leaf was sawn off again at the shoot as it reappeared. So sheep went hungry. In the hot dry areas with light soils, the surface would shift in the autumn with no carpet to hold it in place. Gold miners helped transport them round the province by releasing pairs near the diggings to breed fresh meat. That's the way they got over the Clutha River.

In 1876 a government committee of inquiry was set up as a result and the same year the Rabbit Nuisance Act was passed giving responsibility for control to local organisations. It didn't work because some farmers weren't as badly affected as others and the rabbit had a refuge if they didn't try to hunt him out. It was a bad year south of the Waitaki, as Burdon pointed out:

> *In 1876 on one run where three years before hardly a rabbit could be seen, sixteen men and one hundred and twenty dogs were employed in keeping them down. They killed 36,000 in one year; but the numbers did not seem to be diminishing. On another run 50,000 had been killed; but the average weekly expenses of £27 were fast ruining the owner.*
>
> *On the plains men hunted them down with dogs; on hilly or broken ground, they shot them. Some went in for trapping and others tried poisoning, but this was soon given up, because the hawks fed upon the carcases and died, and they were looked upon as the rabbits' greatest natural enemy. In parts of Southland, where the increase of lambs had been sixty-five and seventy per cent, it fell to twenty. The yield of wool for the province fell by eight hundred bales.*

Rabbits picked out the warmest and sweetest country and destroyed it, and they spared only that which was damp and unhealthy.

Runholders in some parts of Central Otago walked off their land and effectively turned it over to the rabbits. Much of Central Otago became a disaster area and some of it has never recovered. One of my earliest memories is driving a few miles back from Dunedin one night in the 1930s with my father and a friend on a hunting expedition. They shone a strong light up from the road and the hillside seemed to move as the uncountable rabbits raced for shelter. Rabbit was tasty and tender meat, sold skinned and dressed at low prices in fish shops.

A magazine description of Central Otago, published in 1882, read: 'Hills and gullies that used to present a scene of perfect sylvan beauty, with variegated pasture intermixed with sparkling streams and alpine snowcapped ranges, and literally covered with sheep, now look like a deserted waste, as though some gigantic deluge had swept vegetation off the earth.'

Another formal inquiry in the 1880s resulted in more legislation, this time more determined and centrally directed. But the rabbit spread into Canterbury in numbers, into Marlborough, and then infested parts of the North Island: Wairarapa, Hawke's Bay, Poverty

Bay and the centre of the North Island. He thrived anywhere the land was light and dry and the pasture not too rank.

The problem was nowhere else as bad as in Southland and Central Otago. A measure of control was gained in most areas by the end of the century through the use of poison mostly laid with pollard, a fine bran sifted from flour, but an impediment which persisted where the rabbits were thickest was their commercial value. The Department of Agriculture's assistant chief inspector of stock reported in the closing years of the 1800s that the trapping of rabbits for freezing and export would, in his opinion, become an industry of 'large dimensions'. He wrote:

> *Packing houses have been put up at many railway points, and at these houses the rabbits are packed in cases, twenty-five and thirty in a case, and the cases are then trucked to the freezing works.*
>
> *This freezing industry is a difficult matter to deal with. The rabbits are not caught for the sake of decreasing their numbers, but simply because they realise a profit. What must be guarded against is that rabbits are allowed to be farmed for this special purpose. In some parts of the country I found that very little work was being done during December and January in the way of keeping down rabbits, pending the trapping season opening in February. This was deliberate rabbit-farming and cannot be countenanced.*

It is not hard to understand that a farmer — whose ability to earn money from sheep had been severely curbed by rabbits — would want to get some money back from the enemy. In 1893, seventeen million rabbit skins were exported and earned £139,000, and that same year 100 million pounds weight of frozen rabbit meat was exported. The number of skins declined from that level down to only five million in 1906, but the volume of frozen rabbit meat grew to 226 million pounds that year and stayed at around 190 million pounds until the First World War.

The skin export trade burgeoned again to nearly nine million in 1910, worth £130,000, and stayed around that level until the war. These values were minuscule compared with the export returns for wool, but because of the trade in rabbit skins and meat, the country was unable to get on top of the problem until after the Second World War when the government took legislative steps to decommercialise the rabbit. Not that the problem went away, but it diminished somewhat.

The insidious nature of rabbit infiltration, invasion and domination was captured dramatically by Randal Burdon in his 'Canterbury Under Siege' chapter of *High Country* published in 1938:

> *Between Canterbury and Otago nature has placed a very effective boundary in one of the South Island's greatest rivers. From the three glacier-formed lakes of Ohau, Pukaki and Tekapo emerge three rivers which have their confluence some miles to the east, where they form the Waitaki river. Its swift, icy waters might have been expected to provide a barrier impregnable to creatures so puny as rabbits; but they had crossed large rivers before, and by 1887 had begun their attack on the right wing of Canterbury's main line of defence.*
>
> *They passed over the mountainous country between Lakes Hawea and Ohau, where they divided into two armies. One debouched on to the arid Waitaki plains and made its way eastward along both banks of the Ahuriri; the other, moving round the northern shores of Ohau, crossed the Hopkins and Dobson rivers and threatened to cross the Tasman. Further to the south by some means or other they crossed the Waitaki and appeared on the Waitangi and Tiakatarawa stations; and it became apparent that unless they were stopped the whole of the hill country between the Waitaki and Rangitata rivers would soon be overrun.*

Simultaneously the attack was developing from another direction. Two hundred miles to the north, rabbits were pouring into the Amuri district from the province of Marlborough. Canterbury was beleaguered. The position was so grave that the government, taking a leaf from the book of the Chinese, who built a wall hundreds of miles long to keep out the northern barbarian, put up a wire-netting fence, eighty miles long, to keep out the southern pest. Something of the sort had already been done further north, where the Hurunui rabbit fence, beginning at the sea, ran inland for fifty miles. Caretakers were employed to patrol the fence to destroy any rabbits found within half a mile of it on the

Canterbury side; but in spite of these precautions a few got through and reinforced those that had got in before the fence was made.

Even more eloquently, the unstoppable migration is described by that indefatigable observer of nature in Hawke's Bay, William Herbert Guthrie-Smith, in his 1920s classic *Tutira*:

> *Rabbits turned out in the Wairarapa after a few seasons became a curse, increasing and multiplying until many of the local squatters were eaten out, their stations left desolate, and they themselves ruined. Between that period and the date of their invasion of Tutira the hand of man had laid heavy upon the rabbit.*
>
> *Everywhere also its advance had been retarded by the attacks of harriers, moreporks, and wekas, working not for eight hours a day like man, but for twenty-four. Rabbits nevertheless have forged ahead thinly on a vast front, reaching Tutira almost simultaneously by the hill-top, by the coastal route, and by the line of the human highway. There was no strip of open land from coast to mountain ridge along which during the nineties [1890s] rabbits were not filtering towards Hawke's Bay — towards Tutira. It was an invasion which, however disgusting to the sheep-farmer, was full of interest to the field naturalist.*
>
> *The physiography of Tutira has been described — a series of slopes precipitous to the west, tilted gently to the east, and fissured by ravines. It was a pattern well adapted to illustrate the rabbits' desire to follow a definite direction. The invasion moved from south to north, the ravines on the other hand ran east and west, at right angles, that is, to the line of trek. On the high ranges of the west these natural obstacles to progress were particularly formidable. On many the gorges began within a few chains, sometimes within a few yards, of the ridge-cap. Only rabbits, therefore, moving along the very summit could proceed. All others found their progress barred, discovered themselves on the brinks of precipices, many of which were fifty or sixty feet in height.*
>
> *It was impossible for them to remain for any length of time on barren cliff-edges. By natural enemies, necessity of food-supply,*

*fear of the open, and lack of burrow accommodation, they were eventually forced either once more to the summit or downwards to more accessible country. Thus barred from their selected line of progress many were forced downhill, and reached after a time the great pumiceous trough of the run. Through this, at that time, bleak forbidding country ran east and west the main stock-route of the station, the trail by which sheep were driven to the distant paddocks and mustered homewards for shearing. It varied in width from five to ten yards, and for miles passed through fern-lands bare of grass.*

*Now it might have been anticipated that rabbits thus finding themselves on an open road, a smooth space hedged in by deep bracken, would have been content to accept it without question. Far from this happening, however, the east and west stock-route was used no longer than was necessary to find northward-leading exits. No pig-trail, however narrow, no sheep-track, however overgrown, no fence-line, however little trodden, that branched off northwards, was neglected. Rabbits caught by the contractor and his trappers were found in the extreme northern points of the blind spurs.*

Guthrie-Smith was a rigorous observer and recorder of change in his physical environment, the outstanding amateur naturalist of his time. As he went about his farming work, he noted the arrival and settlement in Hawke's Bay of hundreds of exotic plants and many birds and beasts.

Poisoning and trapping rabbits, often by contractors employed by farmers, kept a measure of control but, of course, it wasn't in the interests of those making money from skins and carcases to let breeding stock numbers fall too low. In 1873, 33,000 rabbit skins were exported, about a million in 1877 and more than nine million in 1882. By 1894 exported skins totalled over seventeen million. So it wasn't until trading in skins and carcases could be stopped that the rabbit threat to sheep farming began to dwindle.

It took a long time, though, to kill the trade in skins. According to Percy Stevens in his book *Sheep*, 134 million rabbit skins were exported from New Zealand during the ten years up to 1949, a figure that will astonish most people who thought the rabbit menace was diminishing sharply during the 1940s. There were those who considered rabbit exports so valuable they should be left to earn money for the country, but the cost far outweighed the gains.

The rabbits severely damaged the light-soil pastures on South Island hills and the subsequent erosion damage was often irreversible. Experiments showed that ten rabbits displaced the grazing of one ewe but were far less valuable to the farmer or the country even on the basis of immediate monetary return.

By the late 1950s, the decommercialisation of the rabbit had reduced it from a scourge to a pest. The campaign was backed by a heavy eradication campaign and a claim, which may have been spurious, that hydatids was widespread among the wild population, a claim that successfully discouraged would-be consumers from eating them. The more intensive stocking of hill country in response to aerial topdressing and the aerial dropping of poisoned baits had an effect. The feral rabbit will never be eradicated and sudden build-ups in population in some areas come and go like a plague. An attempt in 1951 to introduce the myxomatosis virus into New Zealand to decimate the rabbit population failed. The disease had wreaked havoc among Australian rabbits. One reason it didn't take here is believed to be the lack of biting insects.

In July 1997, the government turned down an application to import the rabbit haemorrhagic disease (formerly known as rabbit calicivirus disease), having taken scientific advice on two aspects of such an introduction: its effect on the ecosystem and other species and its effectiveness, or lack thereof, against feral rabbits. What followed was an extraordinary disregard for the law when the haemorrhagic disease was brought in illegally by a group of farmers. The government not only never discovered or charged any person but retrospectively condoned the action. However, a few years later rabbits seemed to have made something of a comeback and the effectiveness of haemorrhagic disease was seen to be short-term.

Rabbits successfully adapted to New Zealand conditions, but over many years attempts — happily unsuccessful — were made to introduce into the wild the Canadian moose and the Indian grey mongoose, as well as kangaroos, alpacas, gnus, chipmunks, raccoons, bandicoots, squirrels and even zebras.

The other major animal pests that survived and thrived are deer, possums, goats and, in some areas, wallabies. As with rabbits, they were introduced with the benign intent of providing game in a country without any land mammals of its own. All three found no natural enemies and caused erosion in areas where they flourished by eating out grasses, young trees and shrubs and defoliating larger plants in areas where their population build-up put pressure on food resources. The government was forced to send in shooters against deer and goats. Deer cullers, living in the bush were romanticised and made famous by the story-telling gift of one of them, Barry Crump whose first novel, *A Good Keen Man*, became an extraordinarily successful book on publication in 1960.

The major animal pest persisting into the early twenty-first century, the brushtail possum, has become at least as an unstoppable invader of regions as the rabbit was. It was introduced from Australia from 1858 onwards to establish a fur trade. A nocturnal animal, it is heavily represented in roadkill, blinded by headlights as it continues its spread throughout the country. It populates forests everywhere, but mainly in the North Island, and in provincial towns invades suburban copses, and may live and breed in the space between ceiling and roof. It defoliates trees, may carry bovine tuberculosis, a scourge of dairy cattle, and is hardy in the face of measures to exterminate it.

At various times, possum skins have had small export booms and the soft fibre of its fur has been woven with Merino wool to produce lightweight and warm clothing; although commercialisation will never impact on the feral population. Large-scale poisoning campaigns against possums, rats, stoats and other introduced animals that prey on native bush and native bird life, often involving air drops of 1080 (sodium fluoroacetate) poison, have drawn protests from those who believe it is indiscriminate and therefore cruel and kills birds and domestic animals. But the alternatives, vastly less effective against predators, are judged by most conservationists to be ineffective.

Two imported animals which proved to be of enormous basic value to farming are the horse and the dog — although there was a period in the twentieth century when wild dogs were so feared as predators that farmers gained the right to shoot on sight any unregistered dogs on their property.

The first dogs came with Māori. The kurī was described by Frenchman Julien Crozet in 1772 as: 'a sort of domesticated fox, quite black or white, very low on the legs, straight ears, thick tail, long body, full jaws, but more pointed than those of a fox, and uttering the same cry; they do not bark like our dogs'.

In pioneering days, dogs were used more to contain sheep in the absence of fences than to drove them. The shepherd more often led his flock with a bellwether. Dog training developed as a work-related pastime, a sport in its highest form.

Dogs arrived again with the sealers and whalers and the pure kurī became extinct. The Scottish shepherds who arrived from the 1840s often brought their dogs with them and these and other imports since have combined to make the New Zealand farm dog.

Just as the shepherd's dog became an asset at mustering time, the wild dogs that lived in the bush became a menace to the flocks, and to this day farmers have an affection for their work dogs and a hatred of the wild predator.

In pioneering days, dogs were used more to contain sheep in the absence of fences than to drove them. The shepherd more often led his flock with a bellwether. Dog training developed as a work-related pastime, a sport in its highest form. Animals like Border collies are intelligent, quick to learn and farmers soon came to understand that dogs moving to an elementary set of commands were a huge asset at droving or mustering sheep.

The horse came with Marsden, and there have been large infusions of blood since then to build up the stock of draught horses, station hacks and, lately, sleek and high-strung racing animals. The horse reached its peak as a work animal in the first quarter of the twentieth century and was replaced by first the traction engine and then tractors and motor cars.

During the oil shocks of the 1970s, when fuel prices soared, some farmers claimed the horse could make an economic comeback as a draught animal if energy costs rose high enough. The factors that will always count against this are the high intake of food by the working horse, more than twice that of a cow, and the time and cost of preparing it for work and rubbing it down afterwards, and of caring for the harness.

Experience dictates that no one can bring in any exotic creatures in the twenty-first century because the dangers for the environment could be at least as devastating as the introduction of rabbits and possums — except that government agencies, after exhaustive research, have brought in what they consider host-specific insect predators from time to time in recent years to control such pests as grass grub. Exotic breeds of farm animals have been introduced in recent decades, usually as semen or embryos, using strict quarantine procedures.

In the age of rapid transport moving hundreds of thousands of people and millions of tonnes of cargo through New Zealand each year, the illegal introduction of animals of the size of possums may be capable of being controlled with some ease compared with insects and, worse, diseases. Biosecurity services now attempt to prevent potentially harmful organisms from coming into the country through border control services based on carefully calculated risk assessment procedures around this country, backed by offshore inspections, surveillance and quarantine techniques and facilities.

CHAPTER NINE

# Cold Comfort

The invention and development of refrigerated shipping was the greatest ever single boost to the New Zealand economy. The lucrative export conduit for meat and dairy products from 1882 lifted the young colony from the despair of economic depression. A new prosperity gradually arose from it as well as the confidence and optimism that expressed themselves in the fifteen prosperous and boisterous years following the election of the 1890 Liberal Government.

The country had undergone explosive expansion during the 1870s, thanks mainly to the massive borrowing for public works and immigration through a plan devised by Julius Vogel, Colonial Treasurer in the administration of William Fox. The government borrowed £20 million in just over a decade. The population almost doubled from 237,249 in 1869 to 463,729 in 1879 — with 32,000 migrants in the peak year, 1874. A works programme saw the construction of 1800 kilometres of railway, hundreds of kilometres of road (including bridges), port facilities and public buildings, 6000 kilometres of telegraph line and an undersea cable to Australia.

This extended infrastructure opened up millions of hectares of land to production, land previously too remote to settle or from which to move produce to distribution centres. Along with the increase in stock numbers and enlarged areas of cropping came fencing, as wire and iron standards became available, giving farmers more control over stock. Export earnings from wool were strong and receipts from gold still fairly buoyant.

But with the 1880s came the country's longest ever economic depression in history as wool prices dipped. In 1860 wool fetched sixteen pence a pound, then eleven pence in 1871, and, after a small boom in 1873, it kept trending down to five pence a pound by 1900. Total export returns for wool stayed about the same during the 1880s but from massive increases in volume.

Gold gleamed less brightly in the rocks and riverbanks of Otago and Westland, and the glowering national debt left by the Vogel plan obscured the future of the brave new colony. For the first time, immigrants became emigrants, those who could afford it leaving for the Australian colonies or the United States, and others asking for government help to do so. Not granted. Those settled on small holdings may have had adequate food but no money as the economy seized up and work on railways and roads became scarce.

Meanwhile the flow of settlers on to the land continued unabated. Stock censuses taken in 1871 and 1881 showed sheep numbers increased from just under ten million to thirteen million and to eighteen million during these hard 1880s. Cattle numbers jumped from just under half a million in 1871 to 700,000 ten years later and to 831,000 by the end of the 1880s. The number of horses almost exactly doubled to 162,000 between 1871 and 1881 and increased again to 211,000 by 1891. But the value of both sheep and cattle depreciated sharply as the local market for meat was met by farms adjacent to towns and cities.

Frustratingly, the population half a world away in Europe was surging. In England it had more than doubled from around fifteen million early in the nineteenth century to thirty-five million at the beginning of the 1880s, without anything like as much growth of food production. The impossibility of transporting fresh perishable food across the world by sea had challenged scientists and engineers in a number of countries for half a century. Americans had successfully shipped meat across the North Atlantic packed in ice, but this was not an option for southern hemisphere countries whose ships had to pass through the tropics.

The concept of mechanical refrigeration arose from some experiments by the great English scientist Michael Faraday, but developments were gradual. In the meantime, entrepreneurs in the countries with large production surpluses were exporting meat, mainly in three forms: preserved by boning and then drying, sealed in tins or jars, or set in barrels of tallow.

More than 200 patents were issued for meat-preserving techniques during the nineteenth century, including methods that involved the immersion in or the injection of chemicals. In Australia, the Dangar Brothers began tinning meat in the 1840s and twenty years later Australian meat in soldered tins was selling well in Britain. Many British consumers baulked at the appearance and palatability of preserved meat, aware from painful experience of the poisoning potential of the product when it went bad.

By 1870, New Zealand meat was being canned with only limited success, so relatively little progress was made with exporting, although technological advances in canning were being made and it was the most common preservation method used in New Zealand. Immersing boned meat in tallow was less favoured than in Australia, despite the fact that preserving birds inside fat was an old, tried and successful Māori method.

Prolific Otago chronicler Alexander Bathgate wrote in 1874: 'The low price of stock caused attention to be turned to meat preserving, and there are now several large establishments in different parts of the province turning out by the ton this wholesome and delicious article of food. The value of such meats exported from the colony in 1872, being mostly from Otago and Canterbury, was over £160,000.' Bathgate said that at the end of the 1860s, cattle had been unsaleable in Otago, but the establishment of the preserving factories had re-established some confidence and 'fat cattle were quoted at £6.10s in the Dunedin market in the month of August 1873'.

Most of the factories had been established at points around the South Island by the New Zealand Meat Preserving Company from about 1870 but, according to historian Randal Burdon in *High Country*, many were closed by 1876 at the end of a flurry of business associated with the Franco-Prussian War. Tinned meat did have brief revivals. Twenty years after refrigerated exports began, the Department of Industries and Commerce reported that the New Zealand Agent-General in London suggested inquiries should be made within the colony on whether preserved tinned meat could be manufactured to supply British army units in South Africa.

The department said it managed to negotiate an order from the Imperial authorities for 100,000 pounds of tinned meat from Mr W.A. Thompson of Woodlands, Southland. 'A very much larger quantity could have been placed,' reported the department, 'but only three offers were received in reply to inquiries, no offers being received from the larger meat-preserving works, who generally replied that their surplus output was already under offer to Australian and London merchants in anticipation of the War Office contracts.' So, in 1902, preserved tinned meat was still good business for certain niche markets. (Huge volumes of oats and hay for horses were also sent to South Africa during the war there.)

New Zealand was curiously slow in its response to the possibilities of refrigeration. Australian engineers had been wrestling with the problem for years. Thomas Mort and James Harrison had built a freezing works in Sydney in 1861 — the first in the world. Mort patented his ammonia absorption refrigeration process in 1867 and in 1873 successfully demonstrated it in Sydney, freezing beef, mutton, fish and poultry. But when they sent their first trial shipment to London in 1876 it failed. The meat was inedible on arrival. The following year the first successful cargo of frozen meat was shipped from Buenos Aires to Le Havre, followed by a successful shipment of forty tons of edible beef and mutton from Sydney to London in 1878.

It may have been slow coming, but this one technological stroke would rescue New Zealand from the prospect of perpetual peasantry, or perhaps a kind of latter-day feudalism. Although Julius Vogel's borrowing had left the country deeply in debt, the infrastructure it bought helped give the economy a major surge as soon as refrigerated ships became a reality because access from farm to freezing works to ports was much easier than it would have been a decade before.

The man who made the first constructive move here was William Soltau Davidson, manager of the New Zealand and Australian Land Company. He persuaded his firm to invest £1000 in a trial shipment aboard the *Dunedin*, a 1200-ton, fast sailing ship, operated by the Albion Shipping Company. A killing shed was set up by the company near Ōamaru and when the *Dunedin* arrived at Port

Chalmers in the spring of 1881, butchers starting killing early each morning. The meat was railed to the ship, frozen between her decks and stacked in the hold.

During the ninety-eight days at sea, the ship's master, Captain Whitson, had to contend with a fire in the sails ignited by the machinery and, when the cold air failed to circulate adequately as the vessel passed through the tropics, he crawled into the hold to cut new air ducts. He was overcome by the cold and would have died there had not the first officer crawled in behind him, fastened a rope to his leg and pulled him free.

When the *Dunedin* arrived in London, she was met and her cargo inspected by Davidson, who had ceremoniously stacked the first carcase aboard the ship in Port Chalmers and then immediately left for London. The £1000 proved worth every penny. The 4908 prime sheep and lamb carcases sold within a fortnight for £8000, more than twice the price they would have fetched in New Zealand. The cargo included 2226 sheep tongues, twenty-two pig carcases and a small consignment of butter.

4908 prime sheep and lamb carcases sold within a fortnight for £8000, more than twice the price they would have fetched in New Zealand. The cargo included 2226 sheep tongues, twenty-two pig carcases and a small consignment of butter.

In June the same year the *Mataura* sailed from Port Chalmers with the second refrigerated cargo, again successfully. A year later seven kegs of butter went with the first frozen shipment of meat from Wellington aboard the *Lady Jocelyn*.

The trade boomed from the beginning. A number of steamships were soon fitted with refrigeration holds and some new vessels were designed and built with refrigeration equipment. It was a feverish time in the shipping business, anyway. During the last third of the nineteenth century world trade doubled in volume and the tonnage of the international merchant fleet about doubled also. It was the beginning of the era of the iron steamship.

What is often forgotten is that refrigeration shipping imposed demands on the Railways Department, which responded remarkably well, at first anyway. Thirty-three refrigerated trucks were rolling by March 1883, with fifty-three more within the following three years. Squabbles arose during the following decade, however, about the number of trucks on the various routes between farming regions and the major ports.

According to Percy Stevens, the biographer of influential Canterbury farmer John Grigg, 'the first man to eat Canterbury lamb in England' was Grigg's son, John Charles Grigg, at Jesus College, Cambridge, along with members of the college rowing eight. Grigg had arranged the shipment aboard the *Dunedin* of three of his Shropshire-cross lambs and two wethers among all the New Zealand and Australia Land Company carcases. Grigg the younger, according to Stevens, collected the lamb, then thawed, cooked and served two cold hindquarters to the oarsmen. They loved it, Grigg is said to have reported, and most had three helpings.

As soon as they heard of the success of the *Dunedin* venture, Grigg the elder and a number of Canterbury runholders with large properties got behind the new trade and formed the Canterbury Frozen Meat Company. Freezing works sprang up around Canterbury, Otago and in those parts of the North Island with large sheep populations. Within ten years of that first voyage, seventeen freezing works were operating around the country with a capacity of 3.5 million carcases a year. By 1911 works numbered thirty-one. Beef and pork were also being slaughtered for export, although trade in pigmeat never developed.

Initially, frozen meat was mutton rather than lamb or beef. The Merino with its dark and stringy carcase became a problem. Now that the meat had become such an important export, the Merino's days were numbered as the predominant New Zealand sheep. The breed was brought in from Australia in large numbers from 1840 to serve as the basic flock sheep, the role it had played in Australia. Later importations brought stud rams from Europe, the United States and Australia. The Merino pickety-picked in its goat-like way at pastures lush and sparse over the hills and valleys of the whole country for more than forty years and the woollen industry had been woven from its silken fleece. But in another forty years it would represent only five per cent of the national flock, almost exclusively in the high country of the South Island.

Not that the solution of cross-breeding was new. It was simply accelerated. Some sheep farmers had been putting Southdown or Leicester rams over Merino ewes since the earliest days to produce a less skittish sheep with a better meat carcase for the urban market, a cross that would do better on the stronger, comparatively rank English grasses. The Merino's favourite food was the fine, feathery native pastures.

During the late eighteenth century, British agriculture had gone through a revolution as enlightened farmers concentrated on breeding and on pasture development as a response to the country's burgeoning population. Although the enclosing of common land had socially distressing consequences for village peasantry, it did allow firm control of stock and led to selective breeding towards higher productivity of sheep and cattle. Fenced paddocks also spread the principle of rotating crops rather than having to leave land fallow every few years to prevent its exhaustion.

In the 1760s Englishman Robert Bakewell became the first to adopt some of the animal breeding methods still used today to enhance inherited qualities he wanted to perpetuate in his animals. Not only were his breeding techniques innovative, his concept of breeding sheep and cattle for meat was new too. He helped create the 'roast beef of England' myth because before his time, the main roles in life for cattle were milch cows or draught animals, and sheep were most valued for their wool not their meat. Bakewell changed the scrawny commons-grazing sheep of his county into the English Leicester, famed for its delicious mutton.

Other farmers in Britain followed his example and bred up their own local stock. This breeding revolution that began in the eighteenth century expanded in the nineteenth, and it had important ramifications for pioneering New Zealand farmers who were nothing if not innovative. A number had imported the Leicester, its spin-off the Border Leicester, and also Lincoln, Romney Marsh, Cotswold and Southdown. Farmers were trying for sheep better adapted to heavy country and the new breeds and crossbreds produced better fleeces of their long-stapled wool than the same breeds produced in Britain. As early as 1874, a Christchurch *Press* report announced that Canterbury farmers had been cross-breeding so extensively over the previous two years that the crosses outnumbered Merinos at the saleyards. The tempo then sped up, driven by the demand for meat for the export trade. By 1892 there were only six million Merinos in the national flock of eighteen million.

Over the following decade the volume of mutton in the London export trade diminished as the crossbred fat lamb took over. Partly this was in response to a marketing difficulty. South American countries were dropping huge quantities of high-quality mutton into London, but they couldn't match the plump and succulent New Zealand lamb which appealed to the English palate.

The heavy-set Lincolns, one of the first breeds fixed during the agricultural revolution in England, arrived in 1840, but the most influential importation was in 1862 by the New Zealand and Australian Land Company. It was used immediately over the Merino to produce successive generations of crossbreds and what became known as the 'colonial half-bred'. Lincolns were the most common sheep in the North Island until the turn of the twentieth century.

The first Romneys were put on pasture at Waiwhetu, in the Hutt Valley, by Alfred Ludlam, in 1852. An English Leicester flock was established in Auckland in 1853 and another in Canterbury in 1865. Border Leicesters were brought in by the New Zealand and Australian Land Company and by George Murray of Taieri in 1859.

South American countries were dropping huge quantities of high-quality mutton into London, but they couldn't match the plump and succulent New Zealand lamb which appealed to the English palate.

A Scottish shepherd, James Little, arrived in Otago in 1863 with twenty-two Romney Marsh ewes and nine rams for the Corriedale estate of Dr George M. Webster. Little decided the tussock grass of the station did not entirely suit the Romney and he also aimed to develop a dual-purpose animal more suited to New Zealand conditions than either the Merino or the long-woolled English breeds, so he set about producing an inbred half-breed, using Romney and Lincoln rams over Merinos, and he called them Corriedales. They were the first New Zealand fixed breed.

The same William Davidson whose foresight and organising acumen later launched the frozen meat trade was farming on behalf of the New Zealand and Australian Land Company at The Levels in South Canterbury in the early 1870s. He studied the breeding techniques of Little and fixed a Corriedale flock there. It was a shrewd adaptation that solved some of the problems with Merinos caused by climate and environment. Corriedales and some unfixed crosses with the heavier English rams over Merino ewes produced better carcases (albeit coarser wool) from the English grasses that were then beginning to carpet New Zealand's grazing areas.

The breed that came to dominate all the others was the Romney. A fastidious mother and good milker, the breeding ewes produced useful strong wool and were ideal for producing fat lambs by Down rams. This cross became the solid economic base of the meat trade with Britain early last century. By the time the Merino had slumped to five per cent of the national flock, in 1921, the Romney outnumbered all the other breeds together.

By the Second World War crossbreds made up more than seventy per cent of the national flock with Romney about fourteen per cent, Merinos and Corriedales around five per cent and all the others less than one per cent. The Romney remained dominant. It had been adapted to New Zealand conditions from the original English Romney Marsh importations. However, declining fertility spread through the national flock in the post-war years, even though scientists helped arrest it by convincing farmers to cull infertile ewes as young as possible through a national breeding programme called Sheeplan.

Probably the major reason for the decline in Romney fertility was the breed society's concentration on conformation and other physical qualities not directly related to actual breed performance and productivity. New Zealand Romneys became narrow in the pelvis with wool on the face and legs, whereas the indications were that a clean face and legs were genetically linked with fertility.

Veteran gun shearer and machine-gun publicist Godfrey Bowen, writing about the evolution of shearing techniques in his book *Wool Away*, said: 'The present-day Romney, with wool to the toe and even on the top lip, is much different from the clear, open-pointed Romney of fifty years ago.' The conformation and performance of the breed was dominated by stud breeders, mostly concentrated in the Manawatū-Rangitīkei region and, even though they had fixed the specifications of the Romney, problems affected their flocks too.

Low fertility, still an issue, especially in the North Island, was a reason for the development of two new breeds, Perendale and Coopworth, both fixed from Romney crosses, and both grew rapidly in popularity. The breeding up of fertile strains within the Romney flock, through Sheeplan, and through the efforts of individual farmers who had become more conscious of the need to select for fertility, enabled the breed to hold its premier position in the national flock.

The character of the wool trade was completely changed by the eclipse of the Merino. New Zealand's major output became coarser wool, much of it for the carpet trade which grew vigorously both overseas and here from the early 1960s. The Romney fleece was valued as the major component of woollen carpets, desirably mixed with a coarser wool such as from Scottish Blackface which was no longer available in New Zealand. By the time the need for its coarse fleece for carpets grew, the disease risk had put up a barrier against its re-importation. But the industry was reprieved from the continuing reliance on imported Blackface wool by Dr F.W. Dry, a research scientist at Massey University, who developed an adequate replacement, the Drysdale breed, more or less by accident, in the early 1960s.

The Drysdale sheep had a gross, hairy, long-stapled fleece, lacking in crimp — the sort of wool Dr Dry and his team at Massey Agricultural College were attempting to eliminate from the Romney — until they realised it was just what the carpet industry wanted.

In the 1920s, wool processors were reporting what was considered an undesirable fault in New Zealand's crossbred clip — too many hairy fibres in the fleece. Back in 1932, the Governor-General, Lord Bledisloe, in a lecture on agricultural research, said: 'Kemp or hair in the fleece, especially of Romney Marsh sheep, is a notable and growing defect in the wool of New Zealand flocks and much research both at Massey College and in England has been directed towards its elimination.'

Dry was a foundation lecturer at Massey and because he had worked in England on the inheritance of fleece colour in Wensleydale sheep and also on the coats of mice, he was asked to research how what they called halo-hairs were inherited and how they could be bred out. By 1940, he had found that the culprit was an N-gene, and in the process fixed the eponymous Drysdale, a breed in which this gene was dominant and whose wool was similar to that of Scottish Blackface and some Asian and Middle Eastern breeds not present in this country but valued by the carpet industry. A Christchurch-based carpet manufacturer had applied without success to bring in Scottish Blackface sheep, so the heavy wool had to be imported for blending with other yarn in New Zealand carpets.

The Drysdale produced slightly more wool on average than the Romney, had a similar lambing percentage, was slightly longer in the leg and higher at the shoulder and was open-faced. Rams were easily recognisable because of horn growth.

The commercial development of the breed began in 1961 when control of Drysdale wool production was vested in two companies which bought rams from accredited breeders and leased them at an annual fee to producers of fleece for the carpet industry.

The Drysdale was a useful animal for a few decades but went out of fashion with woollen carpets and no other reason justified the breed's use, especially because it needed to be shorn twice a year, a costly business. Flocks were largely disbanded.

Scottish Blackface had been brought into New Zealand in early years but had disappeared from the national flock. The breed was not the only one known to have been imported and then to have faded out. Among others were Norfolk and Oxford Down breeds (later reintroduced), the Kerry Hill, Cotswold, Wensleydale, Roscommon, Dartmoor and the Tunis. No doubt some of these are still represented in the gene pool in this country. The two New Zealand breeds adapted from the country's varied stock sources were the Corriedale and the South Suffolk.

Refrigerated shipping had an indirect but positive effect on arable farming practice during the 1880s and 1890s. Among the by-products of freezing works were fertilisers, and this encouraged farmers to sow turnips and rape with fertiliser either before or after a cereal crop. This rotation with the additional cultivation enhanced the fertility of the land, improving wheat and oats crops, and made food available for fattening stock. Thus a mixed farm could operate independently, or the farmer could take surplus store stock from hill country properties after shearing and fatten them for the works. This practice was supplanted in the first two decades of the twentieth century by spring topdressings of superphosphate on both cereal crops and pasture.

Dairying was the boom industry in those early years of last century. Because milk needed more careful handling and more complicated processing than meat, dairy exporting developed more slowly. But, although many settlers fattened lambs on early pastures after they had wrested their farms from the bush, dairying could better support undercapitalised smaller holdings by providing a more consistent cash flow. Wool and meat prices were generally subject to wider fluctuations of price.

A result was that milking cows were put on a lot of land only marginally suitable. Those who made it through to a high enough level of equity in their property aggregated their neighbours' properties and moved into sheep; or perhaps got on to better land as dairying gradually consolidated in the most prolific grass-raising areas where soil fertility was adequate and warm rain fell through the year. Much of the land originally broken in for dairying was later turned over to sheep.

The development of refrigeration was a huge boon to New Zealand farming. Wool was the early export winner, capable of being shipped anywhere in the world without the worry of deterioration, but meat and milk have become the commodities that enabled New Zealand to escape Third World status. Also, freezing technology has been refined over the years, providing sophisticated chilling systems that keep produce very close to fresh. Allied with fast and efficient transport using containerisation, chilling has enabled the export of meat cuts to the top end of international markets and also the delivery of fish and a range of vegetables and fruit in near pristine condition to consumers around the world.

CHAPTER TEN

# Status and Security — Land

Wrenching land from the bush in the North Island was a heart-breaking business, especially for the great majority who had very limited capital. But thousands of families were prepared to endure the work of cutting out their forty or fifty acres for the great ideal of land of one's own. Most of the land was especially suited to dairy farming, but until two decades after refrigeration, butter and cheese were seldom profitable.

The cow cockie's lot was not an easy one during the first fifty years from the 1880s when dairying as an industry first became a possibility. Over the years thousands of men in their grey shirts, or raw-wool singlets, with brown speckled working trousers tucked into knee-high gumboots, bowed by the weight of mortgages, groped through their paddocks well before the sun showed, herding their cows and their wives and kids into the cowshed to start a day on which they would see the sun up and down again leaning against a cow's flank.

Milking was a family occupation. The farmer would be outside most of the day, watching the grass warily for the right degree of growth as he moved his herd round the paddocks, feeding out hay in the winter, turnips or silage from the late summer if he grew enough or could afford it, repairing fences and gates and machinery and milking systems, his face blotched with the sun, the skin on his forearms and the backs of his hands like potatoes baked in their jackets.

Meanwhile, his wife would be inside the house between milkings and if she had a large family, as many did, then her day was even longer and possibly harder.

The poet Eileen Duggan spoke for the farmers' wives:

*She is up in the dark and out in the cowyard.*
*Sometimes a star is left in the air,*
*Through the cold interregnum between night and morning,*
*A timid vice-gerent, but what does she care?*

*Milking and stripping, milking and stripping,*
*She leans in against each cow as she stands.*
*She breathes when it breathes and dreams till the milk dries.*
*Her body is gone. She is only two hands.*

*Her face to its flanks and lulled by the milk-beats,*
*She feels her mind empty itself and grow still.*
*As though the long-borne at last is delivered,*
*A lightness has lifted her thoughts and her will.*

*She has the same ease as a sky after thunder,*
*Sure only of silence and safe, dousing rain,*
*Or a country that watches its enemies leaving,*
*And rests before manning its own forts again.*

*There is something unknown between her in her stillness*
*And the cow standing there in its soft clover steam,*
*One chewing the calm of inherited daisies,*
*And one, even-breathed, on the cud of a dream.*

This was one of the too few gestures towards the work of women in New Zealand's pioneering society. Only the nuclear family with its responsibilities apportioned between men and women the way they were then was capable of developing the small farm and it was the women who held that unit together with stoicism, common sense and extraordinary stamina. Many, if not most, of the dozens of local historians have been women and have become aware of this. They have noted that most people recorded and talked about the work and public lives of their fathers and grandfathers and had to be prompted to talk about their mothers and grandmothers.

Pioneer women produced children with what one historian called 'startling regularity', and many died from child bearing or just exhaustion. They received few accolades outside their families for their courage and enterprise.

For many impecunious men and women who chose the wrong land at the wrong time and were inadequately prepared by previous knowledge and experience, the discomfort and work were futile, and they never made it through to solvency. Frank S. Anthony, humorist (*Me and Gus*) and novelist (*Follow the Call*), bought land with government assistance after serving in the First World War: fifty acres, £45 an acre, £100 down. *Follow the Call* is based on that experience:

> *Sometimes I'd get a piece of paper and pencil, and work out how much my cows were going to produce in a season, and how much money I would have saved up in a year. I've kept some of those calculations out of curiosity, and they make me laugh even now. I was only hundreds of pounds out. It was a great experience, sitting in that fire-lit bach, building castles and saying to myself, 'My own boss! No one to give me an order, no one to criticise what I do…' My farm was poor and the pasture thin and run out, and as soon as I discovered that the spring was passing me by with just a feeble nod and smile, I switched my hopes on to the summer flush of grass.*
>
> *It was a shock to find that my heifers started to go off in the summer, in spite of the summer flush, but after a day or two of worry I took heart again. I had put in three acres of soft turnips for autumn feed, and pinned my faith on the turnips. I relied on them to pull me up and make my herd milk well through the autumn months. By the end of February the grass began to disappear off my section as if melted away by magic, and if it hadn't been for the three acres of turnips my cows would have starved to death. Instead of making them milk it just kept them alive, and every morning down at the factory I used to take the lids off the cans before drawing under the hoist and let some of the beautiful turnipy stink get away.*

Anthony's hero stayed on for a happy ending. Anthony himself walked off his farm.

The fight against the bush didn't really end until the 1920s, and some historians say the pioneering period was not over until the 1940s. Late in the nineteenth century and early in the twentieth, a factor in land valuation was proximity to a port and/or a railway service

> Jack wanted to be not only as good as his master but to be his own master, and the small farm in the bush was his best chance.

and the state of the access roads. Nevertheless, thousands of people were prepared to try their luck. It was not just a search for egalitarianism. Jack wanted to be not only as good as his master but to be his own master, and the small farm in the bush was his best chance. When the big blocks of land were opened up for ballot after the Land for Settlement Act passed by the Liberal Government in 1894, hundreds of men put their names forward and drew fifty acres or more. Most had only a shallow knowledge or experience of farming, but they did bring many skills to a community. They were builders, engineers, horsemen, saddlers, blacksmiths, farriers, bakers, butchers, rabbiters, soldiers, seamen, and a few schoolteachers and some with experience in running small businesses. Often the one lack was farming experience, but they were useful to each other in many peripheral ways, and their common bond was the belief that owning their own land was the route to success and independence. And often it was.

When the Whakatane Cattle Company's station was broken up and lots balloted in 1896 and 1898 — releasing seven thousand acres of freehold, followed by leases covering twelve thousand acres of Māori land — more than one person from each family usually put their names down for the ballot to increase their chances. Much of the land was covered by scrub, and drainage difficulties had to be attended to, but it was mostly rich loam soil with very few stumps from forest trees to be blasted out.

For most of that decade the settlers tried growing oats, wheat, barley and even sorghum and sugar beet before settling on maize as a main cash crop. Then in 1898 and again in 1900 severe frosts around Christmas time devastated the fields. Local historian Alison B. Heath wrote: 'Every settler with a decent crop of maize was ruined. Blackened stalks filled every paddock and there was nothing left. … Those with cows still had an income, even if it was a daily chore to milk them.' Thus a dairy company was formed and cows became the main source of income.

Part of the vision of the Wakefields was village settlement by yeomen — a vision tinted by nostalgia after so many yeomen of England had passed into grim city factories in search of wages. Many New Zealand settlers remembered the time in England when a man farmed in a small way if he chose to — that was before the enclosures, and before the more ruthless dispersals by the landlords of Scotland. Some of them carried a loathing of landlordism in their bellies.

Until the middle of the eighteenth century much of the land in England was still farmed in open paddocks by cottagers who shared rights to common land, by yeomen who were freeholders, by leaseholders, by 'copyholders', all of them free from eviction as long as they paid their dues. This was part of the intrinsic structure of English society. But it broke down from the mid-eighteenth to the late nineteenth century.

Ownership of land in Britain carried with it political power, which the gentry proceeded to use to wrest the often undocumented legal rights to land from those they considered their lessers. About 4200 Enclosure Acts were passed through Parliament between 1760 and 1845, resulting in ruthless mass evictions from long-held land. The victims became acquiescent farm labourers, or trudged into the cities, or emigrated — some of them to New Zealand. This dispossession seriously deprived ordinary people who also felt deeply that land ownership conferred status as well as security. Writer, historian and philosopher Thomas Carlyle wrote in 1843: 'The land is mother of us all; nourishes, shelters, gladdens, lovingly enriches us all.' From the days of Ancient Greece, the ownership of land gave aristocracy its status.

Many blamed the ills of Victorian society in Britain on the massing together of people in cities as a consequence of the Industrial Revolution. The countryside was considered the ideal, the healthy, the natural habitat of man. The new city-dwellers could trace their troubles back to the day they left their villages and farms. And it was true that large, densely settled cities, a new phenomenon in the world, were dangerous places. The natural increase in population was higher than it had ever been before.

Investment, planning and technology could not keep up with the consequences of rapid urbanisation, and its consequent problems such as food preservation and distribution, clean water reticulation, sewage disposal and the preservation of order. The material and spiritual woes of the world were blamed on the abandonment of the 'natural' life rather than on the disintegration of a villager's tribal moral and cultural mores.

The belief that the remedy for social and mental ills was to return to the bucolic life was a driving force behind the migration that settled New Zealand. It manifested itself in many ways on arrival, including the quarter-acre section, the siting of hospitals on hills on the edge of town and the building of mental asylums in the countryside. Until well after the Second World War, it was basic to public health policy that gardening, working in the open air, was physically, mentally and emotionally healthier than living in paved urban streets. The Governor-General, Lord Bledisloe, gave the Cawthron Lecture at Nelson in 1932, on 'A Conspectus of Recent Agricultural Research'. He opened with: 'Science is sometimes held accountable for many, if not most, of the world's present ills, such as ultra-mechanisation with consequent unemployment, chemical warfare and irreligion. So it is.'

Charles Bathurst, Lord Bledisloe, was regarded as the farmers' Governor-General (1930–35) because of his deep interest in and knowledge of pastoral farming. He was a regular speaker at farming conferences and conventions during his term of office and his knowledge and experience meant he was keenly listened to. His most famous trophy he left to mark his stay was the Bledisloe Cup for rugby competition between New Zealand and Australia, but he also donated awards for gardens featuring native plants and for the best Māori farmers.

Bledisloe was educated at Eton and Oxford, qualified as a lawyer but then studied for three years at the Royal Agricultural College, Cirencester. He was elected to the House of Commons in 1910, was parliamentary secretary to the Ministry of Food towards the end of the First World War, and then parliamentary secretary to the Ministry of Agriculture and Fisheries, 1924–28. He became chairman of the Imperial Grasslands Association in 1928, and from then on his interest in farming was paramount. He quickly attuned to the New Zealand community during the hard times of the depression, arranging for his salary to be reduced by thirty per cent in line with cuts to the salaries of public servants, and taking a focused interest in gardening and farming.

His most extraordinary and generous gesture to New Zealand culture was made in 1932 when he bought the property on which the Treaty of Waitangi was signed and presented it to the nation with the national centennial celebrations eight years ahead. For more than thirty years, the Early Settlers and Historical Association of Wellington held an annual Bledisloe Day.

On his return to England, he became famous for his Red Poll cattle, kept a dairy herd, farmed pigs and grew grain and other crops. He was elected president of the Royal Agricultural Society in 1946 and the following year visited New Zealand in that capacity on a goodwill mission.

Although he continued in his 1932 Cawthron Institute lecture in Nelson to urge 'balance' in attitudes to science and progress, his printed version had a footnote that made an extraordinary point even for that time: 'By a strange irony Science has also contributed to the marked increase of insanity in proportion to population, by intensifying the pace (and consequent nerve strain) of life by its inventions, and by the humane preservation (and incidentally the reproduction) of mentally defective human beings.'

In the earliest New Zealand years, many settlers found that the idyll of migrant dreams required more land development capital than they possessed, and that the economics of numerous small farms dependent on a domestic market in a new community did not make sense. Refrigeration, though, enabled dairy and fat-lamb farmers who were not too burdened with debt to wrest a living from as little as fifty good acres of land. The operative word, though, was 'good'. Land policy was the crucible of New Zealand politics right through the settlement period, until the Second World War.

The Minister of Lands in the 1890s Liberal Government, John McKenzie, was the son of a Scottish tenant farmer, who was raised at the time of the crofter evictions from the land. He emigrated to Otago. He was a conservative in all other matters except land tenure.

He hated landlordism and believed farms should be owned by those who farmed them, although he favoured leasehold over freehold. He was instrumental in breaking up most of the large sheep runs and placing families on smallholdings. A powerful leasehold movement remained in New Zealand for many years until it was defeated, as it was bound to be in a country that was in the Anglo-Saxon tradition.

Once the smallholders were on the land, they nurtured the ideal that theirs was the natural and therefore virtuous way of life. 'Rural life was regarded as essentially wholesome,' says Keith Sinclair in his *History of New Zealand*. 'The morality of townsmen was (and is) suspect. … To farmers the future seemed to lie between socialistic, loafing townies and hard-working, individualistic, freeholding countrymen.'

This credo and its obverse —townies' belief that farmers were all wealthy and politically favoured property-owners — created social and industrial tensions and broke into violence when it was politically exploited (as in 1913 and 1951).

Farmers had enormous political power and were even permitted a form of gerrymandering for many years in the form of a 'Country Quota'. At one time if the President of the Farmers' Union or, more recently, Federated Farmers wanted to see the Prime Minister, the Prime Minister's secretary would possibly say, 'What time suits you?'

That situation lapsed in the early 1970s and was dramatically underscored by the abandonment of awareness of farming's importance when the Labour Government of 1984 pulled their subsidies away. The power had passed on to more disparate large urban groups. Interestingly, since then farmers have simply got on with their job and proved that the country is still dependent economically on their enterprise and skill.

Are New Zealanders still emotionally attached to the land? They are ambivalent. While thousands are moving from the suburbs into downtown apartments, many are buying beach houses, lifestyle blocks and boats to get away from the urban rush. Local markets also suggest that some are returning to home gardening, once, decades ago now, described by a visiting academic as an essential part of New Zealand culture.

Chapter Eleven

# Land Rape? — 'The Prone Logs Never Arise...'

The destruction of the North Island forest was a sad, indiscriminate land rape; or, as Professor G. Jobberns of Canterbury University once said, it was 'the outstanding achievement of our people in the making of the present grassland landscape'. Perhaps it was both.

The first European visitors to this country were awed by the statuesque giants in the great rainforests, as commercially impressed as aesthetically. The kauri, often with twenty metres-plus of clean trunk to the lowest branches, was coveted by shipbuilders for masts and spars, as well as deck and other planking. The Royal Navy got involved early in the trade which grew quickly in the early years of the nineteenth century. Kauri grew only in the north of the North Island. The depredations went deep into the hinterland of Northland and the Coromandel Peninsula.

The Northland forests were ravaged also by out-of-control fires set by gumdiggers and by both Māori and Pākehā farmers involved in cropping to victual ships and to feed the embryonic towns on the northern coasts. With casual profligacy they often destroyed thousands of acres to clear a relatively small arable area. But the final destruction of the forest came later from the systematic assault by settlers who had bought bush and wanted, indeed needed, farmland.

On one hand, successive governments had conservationists fulminating from the early years of Parliament. A report compiled by Sir James Hector, Director of the Geological Survey, from information supplied by provincial superintendents and released in

1874, was alarmist in its assertion that about half the forest area of the North Island had been destroyed by then. Even if this included large areas of scrub and wasteland, which may not have been in heavy bush when the Europeans arrived, it was still an arresting claim.

On the other hand, the government was beginning to pour millions of pounds into roads and railways to take settlers into the bush where their survival depended on their ability to tear down trees and get the land into grass. Such people left conservation worries to those who could afford them.

Poets waxed wordily about 'this desert of logs ... this dead disconsolate ocean', and...

*The prone logs never arise,*
*The erect ones never grow green,*
*Leaves never rustle, the birds went away*
*with the Bush...*

And prose writers prattled, like the woman in *Brave Days* describing a 'bush paradise':

> *There was a great demand for land [in the eighties] and cheap areas from five to two thousand acres were offered by the Manawatu Company. This offer attracted the attention of a widow in Canterbury, who, for eight hard years, had worked at teaching dancing, painting, and many other things in order to educate her three children. The eldest girl was then teaching and needed no help, one child was at school in England, one in Melbourne, refrigeration was operating — her instinct was right — it was land she must have. ...*
>
> *The eighteen-year-old daughter couldn't resist such enthusiasm, and, like babes in the wood, they left for the North Island. In Wellington, instead of a wicked uncle they found very good friends to give advice and guidance. The Atkinson Government had prepared and was ready to open the first Village Settlement (Levin). Our selectors were strongly advised to consider this block ... and were taken in the Public Works' train to see it. Paradise complete, without flaw or serpent, unfolded itself before their astonished gaze. It was a block of solid native bush in all its original glory.*

*Very few New Zealanders have seen such a sight. ... Tall milling timber grew close and straight like a field of giant wheat-stalks. And the undergrowth — the incredible undergrowth! Nature had filled every available inch with something exquisite. Huge tree ferns found space overhead among a riot of trailing vines. Beneath, smaller ferns pressed for space, shoulder high, and underfoot was carpeted thick with miniature ferns and moss, of magic liveliness...*

*A gang of Austrians were at work. They wore coloured handkerchiefs round their heads and called to one another in their musical foreign voices that mingled with the bellbirds and tuis like a rhapsody. The daughter was faint and dazed and overwrought with too large a draught of beauty.*

This description was written from the candyland of hindsight, with the graft of gaining grass glossed over with all its unremembered pains: 'Well, it became necessary, after a while, to farm fairyland, and that was hard and sordid, but the daughter found romance there, and afterwards true happiness, and the mother a generous measure of prosperity. And so all was merry.'

Most of the voices from the age of the bush-breakers belonged to dilettantes whose sweat had never soaked a singlet in the swelter of the summer forest as a saw bit into the time-toughened sinews of the big trees. The bushmen had no bard, unlike the bush itself, unlike the squatter, the drover, the swagger. They were less well educated, had less time than the southerners who left voluminous diaries and letters and even published books. They weren't as literate. Many came from Scandinavia and Yugoslavia and had no English. It was desperately dangerous work and if one poet came close to being their bard, it was much later when Eileen Duggan wrote 'The Ballad of the Bushman':

*Only a fool would eat his heart out so,*
*And I am old enough to have more sense,*
*But lying by the window makes it worse—*
*Look out across that fence!*

*I have to die, they say, so let it come.*
*A clean break is the best I've always said.*
*And better burn than blight for any crop.*
*So I am better dead.*

*My wife says do no fretting over her.*
*She will be soon upon my heels up there.*
*We have been in it ankle-deep at times.*
*She always pulled her share.*

*No son of hers was ever rocked to sleep.*
*She bore me three here in the bush alone.*
*She was as hard as maire in some things,*
*But faithful to the bone.*

*Each day to us was just so many trees.*
*Night was a minute then. Short is the word!*
*Ah who will wake me out of my last sleep?*
*Her clock was the first bird.*

*It is not for my sons I grudge to go.*
*Good lads like them will never come to harm,*
*The neighbours envy us the name they have,*
*But God, if one of them could farm!*

One thing is sure: for most bushmen each day was just so many trees in a harsh regimen of unremitting work. As G.C. Petersen wrote in his essay, *Pioneering the North Island Bush*:

> *It is difficult to realise in these days the price the pioneer bush settlers had to pay to gain their small and often uneconomic holdings. The inner history of the bush settlements is one of hardship, isolation and privation unknown to the settlers of the open country, of men crushed by falling trees, drowned in swollen bush streams, and of men, women and children lost in the limitless bush and sometimes perishing from starvation or exhaustion.*
>
> *Children died because their parents could not afford the cost of bringing a doctor to their isolated homes, and many a desperate man or woman carried an ailing child ten miles and more along muddy bush trails in search of medical help.*

> *The crowning disappointment for many of the Scandinavian settlers [in the Forty-Mile Bush] was the discovery that the forty-acre farms allotted to them were, when cleared and grassed, hopelessly inadequate to support a family.*

The last of the accessible bush held out until well into the twentieth century. In fact, the pioneering phase of New Zealand farming did not come to an end until just after the Second World War. In the 1950s and 1960s, when I was working in the country and later writing feature articles for the *Journal of Agriculture* and *The Weekly News*, I spent much of my time driving through countryside. Most of the smaller farms were still set in stark landscapes — a house with a small, rough-dug vegetable garden, some macrocarpa windbreaks, a few outbuildings, and a house paddock fenced in by sagging wire.

Gradually, after enough years of good returns and diminishing debt, farmers and their wives planted pretty trees, flowers, built white-painted wooden railing fences around house paddocks and then new houses. The land at last looked not only lived in but loved. But the wan bones of old forests still haunted the hillsides of some remote areas where farmers gave up or just hung on.

Some areas had been so thoroughly denuded even by 1897 that the Secretary for Agriculture, John D. Ritchie, regretted on Arbour Day that year that the interest in the planting of trees was not being kept up. 'There are many odd corners about the farm where a few trees might be put in,' he said. 'The want of shelter for stock is much felt in exposed districts, and the great necessity for it is being forced upon stockowners. The indiscriminate clearing of the bush and ploughing of the paddocks has removed the shelter which formerly existed; therefore it is imperative some substitute should be provided.'

Refrigerated shipping created export meat and dairy industries based, especially in the case of dairying, on the intensive farming of smallholdings in areas of regular rainfall. The Liberal Government of the 1890s repurchased large holdings for subdivision and in other ways determined to provide land for intensive farming by people who would work it despite limited capital.

The main regions where forest gave way to farm, forty or fifty or more acres allocated to each family, were Taranaki, Manawatū-Rangitīkei, the

Seventy-Mile Bush that marched from the Wairarapa into Hawke's Bay (soon to become the Forty-Mile Bush), South Auckland and parts of the Waikato and, later on, the King Country. Most of it was rolling to steep. The settlers, although better equipped with iron and steel tools, used land-clearing techniques similar to those of Māori in earlier times. The first step was to choose the right sort of bush, not only for ease of clearance but also for potential fertility. Rimu was to be avoided and while kahikatea and pukatea suggested rich soil they were hard to burn. Tawa land was regarded as the best.

The tragedy is not that the pioneers felled trees but that they had neither the time nor the capital to discriminate and mill the timber. The transformation from bush to a pasture capable of carrying enough stock to support a family took, on even the most accessible land, at least three years. On rougher land some took ten years or longer, moving from cutting and slashing a new patch every winter to burning every summer. Some walked away.

Nothing is as indiscriminate as fire. As one Taranaki pioneer wrote:

> *As the homes were set in clearings in the middle of the bush, the firing season was an anxious one for all, for often a bush fire got out of hand and became uncontrollable, travelling when and where it would. When the fires were burning the whole country was covered with a pall of smoke, and the school children often had an unofficial holiday. Not that they minded that much for they had the greatest fun searching for the little brown trout that had been suffocated by the smoke hanging low over the rivers or had been choked by the ashes that fell into the water.*

The men had to move fast to get cash returns. With a house cow, a few sheep and a garden their families often soon had enough to eat, but few settlers had any money. Often they were dependent on labouring for roadmakers or railway track-layers for their income. They needed to get stock on the land and be on the way towards self-sufficiency before the job disappeared down the line or up the road.

The first pasture after burning was always rich and fast-growing, accelerated by the potash from the burning off. Although cattle were the best for breaking in the stump-studded paddocks, most of the bush farmers chose sheep because they were more economical and they provided a quicker return. Lincoln lambs mostly, or one of the other strong-woolled English breeds with fleeces that weren't too damaged by picking around the rough black stumps. It was a rush for solvency.

Another pioneer wrote:

> *After the burning of the bush came the sowing of the grass seed — not a leisurely passing to and fro across a clear level paddock, but a heart-breaking scramble up hill and down, over logs still hot from the fires, and cindery black. The men returned home almost unrecognisable as white men; often too tired to wash themselves — too tired to eat. ... It was an anxious time, for if the summer rains came and spoilt the burn, it meant a great loss for the farmer; once the leafage of fallen timber drops and becomes a soggy mess on the ground, it is almost impossible to get a successful burn.*

During the first fifty years from the 1890s when dairying as an industry began to become feasible, the burden on wives and children was back-breaking. The nuclear family was probably the only social and labour unit that could have managed the transition from bush to dairy farm. Wives and children helped with farm work including the twice-a-day chore of milking. Poor roads made moving around difficult, especially in winter mud, and it was not uncommon for children to fall asleep in the classroom after early-morning milking and a long trek to reach school. Wives spent wearisome days milking cows, feeding and servicing the family, and attending to their children's education, a chore doubled when their access to lessons was through the Correspondence School.

This was a different life altogether from that lived by the sheep farmers further south. In *New Zealand Now* (1941), one of the centenary books published by the government, Oliver Duff, foundation editor of *The Listener*, wrote: 'the battle of sheep and cows, if it is not quite the battle of two civilisations, is the battle of two social systems. Sheep make gentlemen and cows unmake them. Sheep leave you with clean hands and clean feet, but cows drag your pride into the mud. Sheep leave you free, cows enslave you.'

In later years, when the government's 'country service' policy meant schoolteachers had to spend their first two years of service in rural areas, many married farmers or their sons and helped lift the level of educational performance in rural families. And as roads improved and a school bus network spread, more farmers' children could attend district high schools.

Agricultural science was a neglected area of study for many years as bright children were directed into what were called the professional subjects. The bleak and endless cycles of farm chores in those times encouraged many to get qualifications that would take them to work in town. The drudgery of housework was not relieved until the spread of electricity in the 1920s began to be followed by household appliances.

Some of the rougher country was still unstumped by the 1920s as farmers worked their way, paddock by paddock, towards clearing land enough to plough it where it was reasonably flat, to fertilise it and put down permanent pasture. Steeper land that would never be ploughed took longer to clear. Many farmers who didn't know enough or didn't have the money for the right seed made initial mistakes from which they never financially recovered. Much of the land was never going to be productive enough anyway and should have been left as it was.

Some of the rougher country was still unstumped by the 1920s as farmers worked their way, paddock by paddock, towards clearing land enough to plough it where it was reasonably flat, to fertilise it and put down permanent pasture. Steeper land that would never be ploughed took longer to clear.

E. Bruce Levy, for many years director of the Grasslands Division of the Department of Scientific and Industrial Research, wrote in 1922 that he was in sympathy with the view that a large proportion of such country would, for various considerations, be better preserved in its native state. It seemed inevitable, though, he added, that large areas of hilly bush country must still be cleared for settlement purposes, and thus it was important that it be done using the best methods and by sowing the best pasture.

But the need to wrench farmland from the bush was too urgent for too many people for any grand design to be kept in perspective. Light shallow soil ran down denuded hillsides with rain that surged over low land in its new unimpeded rush to the sea. It is remarkable in hindsight that the erosion damage was not more disastrous than it was in some areas. Intelligent catchment control, some reafforestation and superior, regularly topdressed pastures on the hilltops gradually eased the problem where slopes were not too steep, but large tracts of steep hill country that would have been shunned by farmers in almost any other country in the world were burnt and laboriously grassed.

In only a few years, the grass cover lost out to scrub, or the pasture was unable to hold on the thin soil cover. Hillsides reverted to scrub or slid away, and the cattle and sheep retreated. Wrote a Director-General of Agriculture, Percy Smallfield:

> *I first saw the full implications of deterioration and reversion of surface-sown hill country during a visit to the central western uplands of the North Island in the early 1920s. The condition of two farms stands out vividly in my memory.*
>
> *One was a farm of about eight hundred acres consisting of one large circular hill rising from a narrow flat terrace of a small stream. The whole hill had reverted to manuka and fern, with fuchsia and wineberry in the gullies. The farmer was reduced to milking a herd of about eight dairy cows on the flat land of the stream terrace, instead of managing a flock of about a thousand ewes, which the section would have carried had not reversion to secondary growth robbed him of the twenty years' labour he had put into development work.*
>
> *The second farm was also about eight hundred acres. It had totally reverted to piripiri (hutuwai) which was so prolific that it had covered fences, stumps and logs, as well as completely smothering every vestige of pasture. The farmer, at the time of my visit, was cutting his last few acres of virgin bush to provide grazing for another year or two for the few dairy cows he was milking.*

Even on some of the better country farmers were hanging on to their land by their fingertips. The truth is that hill-country farming in most regions of New Zealand survived and often flourished, only because of aerial topdressing of fertiliser and aerial seed sowing that began after the Second World War.

These developments enabled hill-country farmers to run sheep and cattle and to fatten substantial proportions of their herds for local and export markets. Beef cattle in particular became an economic weapon for the farmer, grazing rank growth to make grass more congenial to sheep. And beef started to make some money as the New Zealander's consumption rose, and as the demand for manufacturing beef in the United States created a market.

CHAPTER TWELVE

# Cows, Cows and More Cows

Cows have been milked in New Zealand for more than 200 years — since Samuel Marsden landed 'Durhams' from New South Wales at the Bay of Islands in December 1814. No matter what one thinks of the tough old disciplinarian, Marsden brought an abundance of the economically important northern temperate-zone animals and plants into New Zealand and gave the new British-style farming a kick-start. The Durhams were early Shorthorns from County Durham.

Milk brought variety and nutrition to the pioneer diet, and the home paddocks of settlers and mission stations soon had their house cows.

The cows were at first milked in the paddock, by hand into a bucket, sometimes with the cow's head tied between two parallel poles driven into the ground as a restraint. Farmers quickly discovered that cows are creatures of routine with their own social pecking order, so they tried to ensure the same people milked the same cows each day in the same order. Next, small sheds were built to provide milkers with shelter from the weather.

None of this was done with any vision of dairy becoming a large-scale export industry, but rather with production of butter and cheese on a domestic basis: for the family, for the nearby town, and a bit later for sale to more distant markets within New Zealand and, ultimately, Australia. Prospects for cheese were brighter given its longer durability before refrigeration.

Milking twice a day was a labour-intensive chore and even large families could cope with no more than about ten cows. When two or three of them made butter and cheese possible, the milk from the day's two milkings was left to stand overnight in flat, wide-mouthed shallow vessels and the cream poured off in the morning ready for churning by hand into butter.

For cheese, where the skill was available, the milk was left to stand longer to ripen. Rennet from calves' stomach linings was added in round iron tubs. A rennet curd would form and most of the whey tipped off. The curd and the remaining whey were then heated and stirred. Next the curd was salted and pressed into moulds or hoops and then pickled in brine. Pressing with weights continued and the cheese was left to ripen. It needed a lot of attention, scraping and rewrapping with cloth two or three times.

So cheese required expertise as well as more work than butter but farmers, or usually their wives, learnt from what practical advice was available and by trial and error.

As the years went by, Taranaki became the predominant producer of cheese, with Waikato concentrating more on butter. Taranaki also developed first and fastest of the dairying provinces; in fact, it stayed ahead until the Second World War when Waikato/Bay of Plenty took over and rapidly surpassed all other regions.

The hard work was rewarding because cheese and butter were useful products in a subsistence economy often built around barter. James Cowan wrote of his childhood in the southern Waikato:

> *I came to know later how short cash often was, and how settler and storekeeper often had to resort to the barter system in which no money passed. Later on I carried to the township every Saturday on the saddle in front of me a box of home-churned butter that surpassed in excellence of flavour any factory butter of today. We got fourpence a pound for it, not in cash, but took it out in groceries — tea and sugar.*

According to one account from early Taranaki:

> *In most cases the butter ... was traded by the housewives for groceries and other goods from the local shopkeeper. Cash was rarely used. The storekeeper then sold the butter, preserved in salt or brine, in New Plymouth for prices ranging from fourpence to sixpence a pound. Australia was the only overseas*

*market for Taranaki butter in 1880, but this was far from being a satisfactory outlet because of the deterioration which took place on the long journey.*

But, anyway, local markets expanded fairly dramatically during the 1870s and 1880s as the population doubled. Only a small proportion of immigrants went straight on to the land. Many were living in towns of various sizes, working in service industries. Indeed, the expanding market in New Zealand's larger towns attracted dairy imports from Victoria and even Europe. And, surprisingly, exports of butter from New Zealand in 1880, two years before refrigerated shipping transformed the trade, amounted to 2717 hundredweight worth £8350. The butter was heavily salted to help preserve it.

Exports of cheese weighed 717 hundredweight, worth £1983. Most of this was from Taranaki to Australia, a straight run across the Tasman from west coast ports. Dairy farmers near seaports also did business when ships called and needed to replenish food supplies with heavily salted butter as well as meat.

The first true dairying area in this country was Banks Peninsula, where butter and cheese were made in the 1840s for shipment to whaling stations and to Wellington, and it was exported during the 1850s to Australia where hot demand came from the Victorian goldfields. Figures issued by the Registrar-General's office in 1904 gave exports of butter in 1854 as 807 hundredweight worth £7399 and cheese as 169 hundredweight worth £975.

By 1860, the volume of butter was about the same, but cheese exports had risen to 810 hundredweight worth £3535. Most of this would have come from Banks Peninsula. A.H. Ward in his history of dairy companies, *A Command of Cooperatives*, quotes one early commentator as saying eighteen dairies were operating around Akaroa in 1857, each averaging four tons of cheese a year.

Meat was the first big winner from refrigerated shipping. The large sheep farms were run by sophisticated operators in regions that were often plains, well developed with reasonable roads and tracks and open country across which sheep could be driven to freezing works. Butchery was a simple trade and breeding for mutton and lamb was already under way.

The manufacture of butter and cheese was more complicated, required capital to develop beyond the farm kitchen and very few people had the skill or adequate facilities to produce a consistent, quality product in commercial volumes. Without this expertise and an infrastructure, a substantial export trade in the early years was impossible, and without an export trade, manufacture on a factory scale was generally unprofitable. The land most suitable for dairy farming was in high rainfall areas and, especially in Taranaki and Northland, roads turned to mud in spring and summer rains made the movement of milk to factories difficult, sometimes impossible.

About the time the *Dunedin* successfully opened up refrigerated trade, a small number of butter factories were built, mostly by private entrepreneurs, and more were planned in anticipation of export possibilities. But it took more than twenty years for a dairy industry to develop that had the consistently high standards of factory production required to compete successfully against traditional European manufacturers, particularly the English, the Irish, the Dutch and the Danes.

The Danes held a large slice of the British market and it was against their products that the best measured themselves. The Europeans were so experienced in making and packaging their produce that Irish butter and English cheese were imported into New Zealand in the 1860s and 1870s. It arrived in good condition and was popular for those who could afford it. In 1868, dairy imports, mainly cheese, totalled more than 5000 hundredweight and from then until 1881 Victoria dominated the trade providing more than seventy-five per cent of New Zealand's imports.

Probably the first dairy factory, in the sense that it processed the produce of several herds, and certainly the first co-operative, was founded in Dunedin in August 1871 by eight men in the Otago Peninsula home of John Mathieson. At the foundation meeting 'an agreement was made with J. McGregor to be manager of the factory to receive milk, mark it distinctly and to give each his share of whey and to make cheese to the best of his ability … the company to find a person to teach him the art of cheese-making and he agreeing not to teach anyone out of the company without the company's consent'. This factory was set up on Mathieson's farm and a stone storehouse was built (which is still standing). A few years later, it became the Peninsula Cheesemaking Company and later again the Pioneer Butter Company. It established the co-operative principle in New Zealand, but privately owned factories predominated for another forty years.

The company continued to have an impact on the industry nationally far beyond its direct effect. Twenty years later, the government's Chief Dairy Instructor said that on the peninsula 'dairying is made a specialty and is the staple industry of the community' and he submitted to a select committee of the House of Representatives its articles of association as exemplary for other co-operative companies.

Industrial, rural and retail co-operatives were fashionable in the early years of the twentieth century when the dairy industry was developing, and not only in New Zealand but in other western countries. The concept was ideally suited to dairy manufacturing and worked well for farmers right through the century.

However, demutualisation with the issue of tradable shares was raised on occasions as a possible change in structure. In many cases advocates were ideologues who had philosophical problems with what they saw as a form of syndicalism — except that the plant was not owned by workers but suppliers. Such a move was also raised, though, by some more pragmatic farmers who realised that, as the cost of land escalated, an increasing number of farms were owned not by nuclear families or family corporates but by groups of people, many of them investors.

About 11,000 farms supplied Fonterra five years after it was formed in October 2001, down from 13,000 as herds became bigger; but as shareholders, or shareholding entities, numbered around 10,000, multiple farm ownership was obvious. Also, it was known that some farmers, or investors, owned more than one of those entities. Although Fonterra said it had no way of knowing the extent of multiple farm ownership, one estimate by an industry leader was that a third of the supplying farms were owned by about 100 farmers, a trend, he said, that was increasing.

The question sometimes asked was whether investors would need bigger returns than they could pull from the company payout. The worry might also emerge as the company expanded globally that local suppliers' interest would be submerged by expansion, especially if other countries could provide milk in large volumes cheaper than New Zealand. Some observers felt that if the industry became an expanding international consumer business it might need to pull in equity from outside.

But the prevailing feeling in the industry was that the long, strong co-operative tradition had served New Zealand well, and the question of demutualisation was quietened at least for the time being by Fonterra's capital appreciation in the early years of its formation. Also, most farmers worried that a concomitant of demutualisation would be a loss of market power for them, and that short-term financial advantage would lead to lower returns in the long run.

In 1881, the government offered £500 for the first factory to export twenty-five tons of butter or fifty tons of cheese. The prize was picked up by the newly established Edendale factory in Southland, a privately owned operation set up by the New Zealand and Australian Land Company. It was controlled by the entrepreneurial duo behind the inaugural refrigerated shipping trade: William Soltau Davidson and Thomas Brydone.

The factory cost £2500 to build and more was spent on it later. It was set up because its principals believed Southland was more suited to dairy farming than grain growing or the breeding or fattening of stock. Local farmers, however, did not seem to agree, Brydone said later, so the company had to buy 350 milch cows of its own for supply. That was in 1880. They made only cheese at first which they sold for good prices within New Zealand and Australia and, later, less successfully, to London.

Once local farmers, persuaded by results, recognised the benefits and profitability of dairying, the company leased them farms from Edendale Estate land and sold them cows at a price that enabled them to get started despite their limited capital. Soon these farmers were supplying all the milk required by the factory.

Overtones of social altruism accompanied the project. Brydone said later the company paid slightly more for milk than factories elsewhere because, 'We prefer … to get our rents paid regularly, and see the farmers succeeding, to getting big interest, or even any interest at all, on the capital in the factory.'

In 1890, Brydone told a parliamentary select committee the company had 'the very best appliances' for making butter but could not find enough butter-making experts 'even if we have men who seem to take every care and who are as good, I believe, as it is possible to get in the colonies.' This worry about expertise at factory level persisted for years.

Several more factories capable of manufacturing both butter and cheese were set up in the early 1880s. Canterbury, Otago and Southland were still the best organised farming centres and dairying attracted interest; but, as tensions with Māori subsided in the North Island and appropriated Māori land was allocated to aspiring farmers, the bush was being rapidly razed and dairying was starting to explode in the areas where it flourished most for most of a century — mainly in the high-rainfall regions: Taranaki, the Waikato, the Bay of Plenty and Northland, and also in the West Coast, Manawatū and the Wairarapa.

By 1885 butter exports totalled 24,923 hundredweight (worth £102,387) and cheese 15,245 cwt (£35,742). Twelve years later, after the spread of the centrifugal separator (introduced in 1880), and the more general use of refrigerated shipping, butter exports were up to 75,287 cwt (£297,515) and cheese to 71,662 cwt (£135,711).

During that twelve years, the understanding grew that dairying would need to become a factory manufacturing industry to gain a consistent quality of product for the required volume, especially with butter. As dairy produce moved into the yawning British market greater evenness of quality was essential.

The 1890s was a watershed decade for dairying. A select committee of the House of Representatives discussed and reported on the industry in September 1890. It noted the value of exports of butter had increased during the decade from £8350 to £146,840, and of cheese from £1983 to £67,105. Despite some optimism these figures inspired, the committee considered the export trade was in an unsatisfactory state, that butter was 'frequently made in a faulty manner' and even good butter was often spoiled in transit 'through defective carrying and transhipment arrangements or the use of unsuitable packages'.

The many experts who appeared before the committee gave conflicting opinions on the cause of both butter and cheese arriving in London in poor condition — including on whether or not it was produce made badly here through ignorance; whether freezing did or did not damage butter; whether tawa kegs tainted produce and tōtara did not; whether shipping delays caused produce to be left waiting in bad storage conditions; and whether shipping companies provided space as suitable for dairy produce as they did for meat and yet charged more for it.

In its report, the select committee recommended the government should give every attention to developing dairying because it believed the market in England to be 'practically unlimited' and European countries had nearly reached the limit of their production.

Inspectors were appointed in London to check on quality and experts were brought to New Zealand from Denmark, Canada and Britain. An experienced Scottish dairy instructor and organiser appointed Chief Dairy Expert by the government the year before the Select Committee hearings, John Sawers, was already having a heavy impact on the industry. His first job was to give practical demonstrations and lectures on cheese and butter making at the South Seas Exhibition in Dunedin. He said the factory system was essential for the future, advocated a co-operative system and his conclusion was the industry had problems that sprang from 'the want of knowledge of many of those engaged in the manufacture and from the want of co-operation in the interests of the factory on the part of the settlers and milk-suppliers'.

A blunt man, Sawers made enemies in the industry but was so effective he became the first Chief Dairy Instructor when the Department of Agriculture was formed in 1892. It was later said of him that his energy and imagination were instrumental in the evolution of a well-organised co-operative factory system.

By 1890, there were about 150 dairy factories in the country, only a shade over forty per cent of them co-operatives. But, as Sawers later pointed out, problems of quality control persisted despite government intervention with controlling legislation in 1892 and again in 1894.

A government dairy produce grader, Mr A.A. Thornton, reported to Parliament in 1897 that it was a 'a great pity' the Dairy Industries Act 1894 did not empower inspectors to prevent the shipment of the 'most inferior quality' of butter. He described one consignment of butter arriving in London as 'a perfect disgrace', and a consignment of cheese as 'weak in body and texture and defective in flavour'.

A government dairy produce grader, Mr A.A. Thornton, described one consignment of butter arriving in London as 'a perfect disgrace', and a consignment of cheese as 'weak in body and texture and defective in flavour'.

The first Taranaki dairy factories, both privately owned and co-operatives, had opened during the 1880s, and a key entrepreneur at the beginning was a Chinese storekeeper, Chew Chong, who had migrated to New Zealand during the gold-rush period. In the Waikato, the butter pioneer was a Cornishman, Henry Reynolds, who established the Anchor brand with the first factory in that region. Reynolds expanded within the Waikato and then into Taranaki, but, like Chew Chong, was swallowed up by the co-operative movement during the 1890s. His company eventually became the huge New Zealand Co-operative Dairy Company which continued with Anchor butter.

The co-operative movement helped shape a generally collective attitude towards export marketing. Although Chew Chong's exports to China of 'Taranaki wool', the edible wood-ear fungus that grew on decaying trees around the region's newly cleared pastures, exceeded exports of butter and cheese through the 1880s, the overseas market for dairy products gradually expanded and was worth £197,000 by 1887, and £350,000 just six years later. During the period to the First World War, individual dairy companies did their own selling through agents in Tooley Street, London's butter market. The industry was moving towards a fully collective marketing organisation when war broke out and the British and New Zealand governments came to an agreement that developed into the purchase of all New Zealand exports as part of the war effort. Demand meant cheese production grew quickly because it was such a useful wartime food.

The bulk buying agreement ended in 1921 when producers went back to competitive selling — but only for long enough for the industry to realise this method dragged prices down. A group of seventy co-operative factories formed a limited liability company to market their produce jointly in London. By 1926, the company had ninety members. In the meantime, the government established the New Zealand Dairy Produce Export Control Board (progenitor of the New Zealand Dairy Board) and in 1926 it took control of exporting including prices — only to run into stout opposition from free traders in Britain.

So the price-setting had to end, but the concept of orderly marketing by co-operatives banding together remained. The industry arranged freight and insurance contracts, regulated shipping to ensure continuity of supply, and introduced a national brand, Fernleaf.

Two New Zealand companies — Amalgamated Dairies Ltd and Empire Dairies Ltd — acted as wholesalers on Tooley Street, but in 1936 a new Labour government installed a guaranteed pricing scheme by which it bought all the butter and cheese from producers at a fixed price.

The plan was to smooth out returns to farmers but, although it lasted for twenty years, the industry was never happy with the pricing mechanism. A provision existed for government to support the industry fund when it went into debit, but it was almost always in credit and when subsidies were tugged from underneath the industry in the mid-1980s, dairy farmers landed more softly than otherwise would have been the case.

But in the depth of the Depression, the industry was in a graver crisis than any time before or since. A survey in 1934 revealed that about half of New Zealand's dairy farmers were technically insolvent. A national scheme to help them refinance mortgages on a long-term basis at low interest staved off total failure until a pick-up in the world economy and dramatic intervention by a new government in 1936 resulted in a gradual recovery. During the Second World War Britain again became the sole buyer of all surplus dairy production in New Zealand, an arrangement that lasted until 1954.

In the early days of the industry, small isolated herds in the North Island made transport to factories difficult. An answer was a proliferation of district creameries or skimming stations to which farmers could take their milk, have it separated by machines, take their skim milk and return home. The creameries delivered the cream to the nearest factory. Mitigating against quality produce were the idiosyncrasies of individual farmers.

The standard of hygiene was not always high either in cowsheds where milk was stored sometimes for two days in the summer heat at the height of the season. For many years, the chief complaint against New Zealand butter in Britain was its 'fishiness', a smell and taste that resulted from cream that had gone off before the butter was made. The creameries were, nevertheless, a necessary convenience for remote farmers, some of whom were so isolated by bad roads, hills and streams they had to transport their milk in cans on packhorses.

The spread of the home separator and improvements to roads brought the demise of the skimming stations. With the reduced burden of only cream rather than full milk, each farmer began the day with a visit to the district dairy factory — their dairy factory. After a slow start, the concept of the co-operative dairy company spread like a faith, an extension of the smallholders' desire for as tight a mastery as possible over their own destiny; and, of course, it gave them a higher return from the product their raw material was processed into.

While the meat industry in most areas was dominated by relatively few runholders who were powerful enough to buy shares in private freezing companies if they wanted to, the dairy supplier was almost invariably a smallholder with limited capital. (Two-thirds of a total of 929 holdings in Taranaki County in 1881 were between ten and two hundred acres.)

Dairy co-operatives were a thing of the times, and in other countries as well, including the United States. Although co-operatives in New Zealand were still outnumbered in 1900 by 152 proprietary factories to 111, the movement was unstoppable and within a few years the proprietary dairy factory was virtually extinct.

In 1902, the secretary of the newly formed Department of Industries and Commerce, T.E. Donne, advocated in his annual report that freezing works should be run as co-operatives and 'in some instances State control is strongly advocated. Whatever may be the merits of these different schemes there can be no doubt that the distribution and selling of frozen meat should be put on a better footing than it is at present.'

Although some powerful meat company co-operatives did emerge, the ideal never arose as anything like as powerful a movement as it did among dairy farmers. Practical and financial reasons helped drive the dairy co-operatives movement. As exports grew from £208,000 in 1890 to £1 million in 1900, British importing houses began to provide loans for capital and felt more secure where the suppliers were involved. However, the importance of privately owned factories in the early days, especially in Taranaki and the Waikato, should not be undervalued.

As roads improved and especially after tanker collection spread throughout the country, factories tended to become centralised and bigger. By the end of the First World War, dairy factories around the country numbered 638 and cows totalled about 750,000. Cow numbers continued to rise to about two million in 1960, by when manufacturing plants had dwindled to 268.

Then the rush of amalgamations began with the aim of gaining economies of scale. During the following twenty years, the tanker became ubiquitous, picking up wholemilk from even the most distant farms and delivering to bigger and bigger and more central factories.

As herd sizes climbed from an average of sixty-five cows to 122, with sealed roads becoming the norm, amalgamations reduced factories to 105 by 1980. Output per factory more than doubled.

As the factories got bigger, the standards of hygiene and quality control improved. Butter making in particular became mechanised with the cream poured in one end of the operation and packages coming out the other. Cheese making became mechanised too where the output of a single type was large enough in volume.

Dairying developed over the first sixty years of the twentieth century as a compact industry with factories centred on intensively farmed grassland of the highest quality, and with research, production and marketing co-ordinated by producer organisations in their various incarnations, by the Department of Agriculture, by dominant manufacturing co-operatives, and by a number of research and regulatory bodies. Government-sponsored research was predominantly empirical, and the progress was striking between the 1920s and the 1960s when Britain contemplated joining the European Common Market.

But as early as 1920, the importance of rich and sturdy permanent pasture types, and the need for lime and superphosphate, was fully understood, and the best farmers were grazing their paddocks in rudimentary rotational systems, making silage for winter feed. There was more emphasis on fodder crops, but the basic approach to pastoral farming was settled. Developments later were mainly refinements, many of them made possible by technological advances in farm machinery — electric fences, tractors, trucks — the whole range of appliances and machines made possible by the wider use of electricity and the internal combustion engine.

Perhaps the most influential man on pastoral farming in his time, Dr C.P. McMeekan, the superintendent of the Ruakura Animal Research Centre, wrote in 1955: 'Thirty years ago our industry was already a long way along the road. The turn of the century had seen its emergence from primitive conditions and methods. Refrigeration had opened up the markets of the world. The internal combustion engine had revolutionised transport, land development and milking methods. … Thus the first twenty-five years of the century saw a remarkable tenfold increase in the exportable surplus of butter.'

Many of the rituals changed, but the basic practice of livestock farming didn't alter that much. It was more of the same done better. McMeekan also wrote of the major changes in techniques of dairy farm management and animal husbandry in the previous thirty years:

> *The concept of controlled grazing as the logical and most efficient way of using pasture has found its applications on every farm. The electric fence has been a powerful force in this direction by cheapening sub-division costs and increasing the efficiency of control.*
>
> *New types and strains of pasture plant have become available together with the necessary knowledge for their establishment and maintenance for maximum yield. Top-dressing with phosphates is universal, while lime, potash, nitrogen and trace elements are widely used where necessary with spectacular effects on productivity.*
>
> *Special techniques for pasture renovation and improvement have been developed. In favourable localities, the menace of dry*

*summers has been removed by spray irrigation. Hormone and other weed killers have banished the bug-bear of most noxious weeds. All these have combined to increase both total yield and total annual spread of grass growth.*

*The concept of fodder transfer from periods of plenty for use during periods of scarcity through saved pasture and through hay and silage has become widely recognised as a key tool for high production. Making hay, silage or both is the pre-occupation of every efficient dairy farmer today in the late spring and early summer months.*

*On the breeding side, the principle of progeny testing as a basis to national herd improvement is universally accepted. Bulls of proven merit or their sons from proven cows head the herds of dairymen seeking progressive improvement in their cattle.*

The scientists' word had spread effectively. In 1921 the average butterfat per cow was 174 pounds, with about one million cows in about 50,000 herds averaging nineteen cows. Exactly fifty years later just over twice as many cows (2.2 million) averaged 274 pounds of fat on fewer than half as many farms (21,900) in herds nearly five times as big (95 cows).

Average fat production had not climbed markedly in the thirty years after 1941 when it got above 260 pounds for the first time. It passed 300 pounds several times in the good years after 1965 but dropped to around 260 pounds on one occasion too. As it was in 1921, the best-managed herds were far and away above the average.

Change during the first seventy years of the twentieth century seemed inexorable but was steady. Then Britain joined the European Economic Community and ten years later New Zealand deregulated her economy, opening up her borders to the full blast of market competition. The quickening of world trade abruptly ended the pastorale of New Zealand dairy farming, the hard but satisfying life through the recurring tenor of the seasons.

## CHAPTER THIRTEEN

# Milk by the Billions of Litres

The dairy industry, fully recovered from the pessimism that shrouded the future in the 1970s and 1980s, entered the twenty-first century shaping itself for leadership in global export markets. Cows numbered 3.9 million in 12,000 herds producing 14.8 billion litres of milk, ninety-five per cent of which was exported in one processed form or another: milk powders, cream products (butter and others), cheeses, and milk and whey proteins. These products were New Zealand's biggest single export earners.

International prices would continue to rise and fall, subject to consumer demand and to the distortions of subsidies of dairying in developed countries, but optimism was abroad as farmers invested in their industry and new farms crept beyond the fringes of traditional dairying country. But for all that, more than eighty per cent of dairy cows were grazing the lush pastures of the North Island — more than a third of them (1.6 million) in the Waikato-South Auckland-Bay of Plenty region. Taranaki was next with 530,000. In the South Island, though, Canterbury (452,000) and Southland (357,000), where land was cheaper, were advancing.

It had been a long, hard haul through such crises as Britain joining the European Common Market in 1973 and the deregulation of the New Zealand economy in the following decade. Soon after the Second World War, butter and cheddar cheese export sales began to contract. Britain traditionally absorbed more than eighty per cent of the world butter trade, but demand in that market was stagnant as the fifteen-year-old New Zealand bulk export agreement with Britain expired in 1954. About the same time, the international market was being corrupted as European countries expanded dairy production, supported by subsidies, from 830,000 tonnes to 930,000 between 1954 and 1957, the year in which the European Economic Community (EEC) was formed.

Europe was recovering from the war and under the EEC regime retreated behind import barriers. Next came the spectre of Britain also disappearing behind this European protectionist fence and stimulating even greater overproduction. All the while the United States was dumping its subsidy-promoted surpluses on world markets, setting international prices in a nosedive. Against this calamitous backdrop, the output of New Zealand butter and cheddar cheese was growing.

Wanted were new buyers in new countries for milk products. The Dairy Board, responsible for international marketing, shrewdly dispatched technical staff into Asian, Middle Eastern and Latin American countries to set up recombining plants in which products with local demand could be manufactured from imported New Zealand milk powder. The board had acquired some expertise in manufacturing offshore. A recombining plant had been set up in the West Indies in the 1950s to meet local demand, mainly because that was where cargo ships from New Zealand stopped for coaling on their way to Britain. It was a small operation but successful and had given technical staff knowledge and experience that proved useful two decades later.

A plant was opened in Singapore in 1961 followed by others in South East Asia and Mauritius. Skim-milk powder and casein, by-products of butter making, became a major export. Adding nutrition to stock feed, they were used for the manufacture of sweets and beverages, bread and other bakery products, and even sausages. In many cases the recombining plants were placed in local ownership with New Zealand contracting to supply the raw material and to provide technical expertise. Markets for spray-dried milk powder spread during the 1960s and 1970s to Japan, Arab states and South and Central America.

The drive for a wider range of consumer products from milk forced up the tempo of amalgamations of co-operative companies in New Zealand. Farmers accepted that the capital demands for modern machinery rendered small dairy factories run by home-town or even provincial companies impossible in the context of world trade with a shifting demand for various milk products. Small co-operatives with ageing plants could no longer offer their member-farmers a payout that would match the amounts available to those belonging to expanding companies investing in large, efficient, flexible plant.

In some cases, resistance to amalgamation arose among communities which had nostalgic loyalty to their local co-operative with allegiances going back several generations. The original factories had been spaced according to how far a horse and cart could carry milk over not very good roads once a day. Some of these small companies wanted to stay separate long after it made sense to do so. If they tried to increase flexibility of production in their own small area, they tended to overcapitalise and fell behind in payout levels anyway. Gradually, economic reality always prevailed. The basic commodity remained cows' milk but, increasingly, manufacturing capacity needed to be large and flexible to shift its production of milk's derivatives — butter, a myriad of cheeses, milk powders, caseins and a range of other specialist products — in response to demand among dozens of markets.

Since the 1960s, roads had been upgraded throughout the country enabling the big tankers to pick up milk on spider-web collection runs. In 1961, dairy factories numbered more than 165 but by 1973 had dropped to seventy-five. In 1971 Taranaki was home to forty companies. Twenty years later there was one.

The next big challenge, thirteen years after Britain joined the EEC, was the deregulation of the New Zealand economy. The comforting rug of subsidies with its intricate pattern of payments was pulled out from under farming over two seasons. During that period, the Dairy Board legislation was amended nine times as farming politicians scurried back and forth to the Beehive. Farmers with high debt against their properties were soon in crisis and some foundered. Dairy farmers had advantages, though, over meat and wool growers: the industry account that supported prices was in credit; most dairy farmers and their families were milking cows themselves with herd sizes then around 150; they were likely to have less exposure to debt because of the traditional stepping-stone approach to farm purchase, via sharemilking; and they had better cash flow.

After the Northland Co-operative Dairy Company amalgamated with Kiwi, the industry had two dominant companies left in the late 1990s, each based in one of the great national dairying regions: Kiwi in Taranaki and the New Zealand Dairy Group in the Waikato and Bay of Plenty. Only a handful of small companies stayed outside the giant conglomerate, Westland Milk Products and Tatua Co-op Dairy Company, and some set up soon afterwards. Westland Milk Products, with suppliers on the West Coast and Canterbury, failed to thrive and, to the dismay of most New Zealanders, was bought by Chinese-owned Yili Group in 2019. Waikato-based Tatua survived as a manufacturer of specialised food products.

Otherwise, the industry's manufacturing capacity was taken within the Fonterra monolith to achieve economies of scale and prevent the likelihood that two companies would compete for price in export markets and diminish the value of both.

Before that final merger, the concept of a duopoly with Kiwi and New Zealand Dairy Group competing had been held by some to be the ideal structure, one that would foster competition in both the domestic and overseas markets and force the efficiencies that competition brings. The Commerce Commission's first ruling was against a merger, but the government supported the monopoly and approval was finally given on condition the new, big company took some sell-off steps to ensure domestic competition.

Thus Fonterra was born, a huge co-operative company, one of the top ten dairy companies in the world and the biggest exporter (to 140 countries), responsible for one-third of the world's dairy trade. It gave itself a dreamt-up, neutral name (based on the Latin *fons de terra*, meaning 'spring from the land') that bespoke the company's determination to hear no parochial echoes. The rules of engagement for farmer-members were changed. Previously, they entered their co-operatives with minimal fees and were paid out on the volume of milk sold; but variable joining fees were now imposed based partly on the size of the supplier. But for all its size, Fonterra ran into financial trouble in 2019 through failing international investment and sclerotic management and it needed urgent attention to survive as the world's biggest dairy exporter.

As the co-operative dairy companies were getting bigger so were farms and the herds they carried, and the composition of those herds was changing too. The most successful breed in the beginning was the Milking Shorthorn. The first Ayrshire bull arrived with the Otago settlers aboard the *Philip Laing* in 1848. 'Rob Roy', as he was called, had an immediate impact on dairying in that province and Ayrshires have been a positive influence in the national herd since.

The first Ayrshire bull arrived with the Otago settlers aboard the *Philip Laing* in 1848. 'Rob Roy', as he was called, had an immediate impact on dairying in that province and Ayrshires have been a positive influence in the national herd since.

Another early importation was the Dexter, also known as the Kerry, but it disappeared. Two breeds have dominated the industry — Jerseys and Friesians (Holsteins). The first Jerseys arrived in 1862, brought in from Jersey by Thomas Syers of Wanganui. Other importations followed and it was in Taranaki that the merit of the small-framed, fawn-coloured cow with its rich lode of butterfat was first recognised. Herd testing in the Waikato underlined the extraordinary productivity of the animal. By the middle of last century, Jerseys numbered 1.5 million, Friesians about 110,000, slightly more than Milking Shorthorns.

Jerseys received the imprimatur of government researchers immediately after the Second World War after prolonged testing. But this was a time when payout was dominated by butterfat content for the making of export butter. As the trend changed to milk volume with other ingredients of milk at least as important, the predominant breed became Friesian. In 1972, Jerseys still comprised fifty-eight per cent of the national dairy herd of about 2,900,000 cattle, Friesians thirty-four per cent, and Ayrshires three per cent.

By the end of that decade Friesians outnumbered Jerseys, and the two breeds have staved off challengers from other dairying breeds brought in since the 1980s. The big-boned black-and-white Friesians gave easily the highest volumes. Friesians were also especially valuable for cross-breeding with beef sires for dairy beef. Gradually they gained the ascendancy. The prevailing view now is that a place exists in the national herd for any breed or cross that is most efficient for a particular productive purpose, but now that butterfat is not the main requirement the big Friesians provide milk volume rich in solids not fat.

But it was herd size that changed mostly in the modern era. The days disappeared when a family could make a good living from thirty or forty cows. Technology paved the way. Milking machines became increasingly efficient. Hand 'stripping' of cows ceased and herringbone sheds proliferated and increased throughput. The pits enabled milkers to avoid back-breaking bending to put on and take off teat-cups. The national herd of dairy cows grew from 370,000 in 1900 to 1,850,000 in 1950, but the average herd size increased only from twenty-plus to thirty-something during that half-century. After that, gradually at first and then very rapidly indeed, the demographics of dairying — or bovigraphics to coin a word — changed as the number of herds declined, growing individually ever larger, and dairy farms began to spread into country traditionally avoided because it lacked adequate rainfall spread widely through the year.

In the thirty years from 1975, the number of herds dropped from more than 18,000 to 12,500. Average herds grew threefold from 100 cows to more than 300. A thousand cows became common and the biggest herds were two or three times bigger than that. The explosion in growth occurred in the last decade of that period with total cows surging from 2.8 million in 1995 to 3.9 million and average herd size from 193 to 315. Average milk solids per cow graphed up well, from 271 kilograms to 308 with some small annual dips caused by unseasonal weather.

Again, technology gave impetus to the trend — the rotary shed, developed by a New Zealand farmer, meant more cows could be milked at a time with less labour, although labour remained the problem it had been for many years. But new urgencies were emerging that would require scientific attention, among them: foot soreness in large herds on big farms having to walk long distances to the shed and losing some condition in the process; declining fertility; and environmental pollution from the intensity of modern dairy farming practice. Between 1995 and 2005, the average number of cows per hectare grew from around 2.5 to 2.8 and that meant more nitrogen into streams, rivers and lakes from fertiliser run-off and from cattle urine and manure.

Dairy farmers were unhappily resigned to incurring costs in the fight to keep the rivers and lakes free as another change in farming practice loomed: milking once a day instead of twice. Many farms had moved to once-a-day (OAD) and more were investigating a change. The advantages were the obvious lifestyle improvement for farmers and families, lower labour costs and less strain on cows getting to and from the shed. Early Livestock Improvement research demonstrated that some cows adjusted to OAD better than others and the company developed a breeding index ranking bulls and cows on their performance under OAD conditions.

Farmers making the change needed to patiently modify the composition of their herds to get the best results. Milk production losses in trials were variable, anywhere between six and eighteen per cent. In some cases, farmers compensated for this by more intensive stocking to get better use of the grass. The advantages of OAD milking were strong enough to drive a firm trend.

The national average of lactations for cows was dropping as the new century dawned with the percentage of cows coming into calf declining. The reasons were not entirely clear, and it was a problem afflicting herds worldwide, perhaps because of the rapid international transfer of genetic material. The situation demonstrated that the focus on genetics for advances in milk volume and quality could result in a distraction from other desirable traits. But the New Zealand industry is so well organised now it can shift focus with relative speed towards breeding for higher fecundity. One approach taken by farmers is cross-breeding for hybrid vigour by using top-quality Jersey semen over their Friesians, thus enhancing in-calf percentages immediately.

In the very big herds now, the cow is a number, a stock unit, where once she was also a name, a personality with a private reputation. The farmer at first built up a herd from affordable local cows and bulls. Compared with the modern cow grazed rotationally, their performance was poor. But the government and industry-sponsored artificial breeding scheme lifted performance.

In the days when a herd of forty or thirty or even only twenty was an economic unit, milking in the early morning and late afternoon was an unremitting family chore, but each cow was named and was known for her behaviour as an individual, her dominance or passivity in the yard and in the bail. Before milk yield was precisely measured, the family knew the generous milkers from the parsimonious, the irritable cow from the tractable; and the good farmer built up a family profile mentally, perhaps, for a small herd, or in a dog-eared notebook that served as a herd book. Farmers culled not only the cows who did not come into calf, but noted those that took more than one service to become pregnant; although even after herd testing and artificial insemination became available they would keep a bull, if they could afford it, for a second chance or for the cows that came in heat late, or they might share a bull with a neighbour.

The industry in New Zealand has depended almost entirely on four cattle breeds: Shorthorn, Ayrshire, Jersey and Holstein-Friesian, all of them, except the Shorthorn, eponymously named for the regions in which they were developed.

The earliest herds were dominated by the Shorthorn, which for a long time was the most important cattle breed in New Zealand — a good milker which also produced excellent steers both for draught work and for meat. Milking Shorthorns populated many dairy herds around the world and a number of successful beef breeds have been fixed from Shorthorn crosses. The breed was developed in England's Durham County from a cross with imported Dutch cattle, and then one breeder selected for milking ability and another for beef.

Shorthorns were still called Durhams when they arrived in New Zealand in 1814 with missionary Samuel Marsden as a gift from New South Wales Governor Lachlan Macquarie. They spread rapidly through the country, serving as multi-purpose animals. As in England, some were selected for beef herds and others for their capacity as milkers. By the 1920s more than half the dairy cattle in New Zealand were Milking Shorthorns, but later that decade they began to give way to the Jersey. By the beginning of the twenty-first century, as the bicentenary of the breed's arrival here approached, so few were left in dairy herds a move was afoot to preserve some animals in recognition of the breed's service to New Zealand farming.

The Ayrshire was developed during the eighteenth century when the rudiments of crossing cattle and selecting for certain traits were becoming understood. Before the breed was set in Scotland's County of Ayr, crosses involved some European strains, perhaps some from the Channel Islands, and certainly Teeswater cattle from Britain's Tees River Valley which were also progenitors of the Shorthorn. On the way to its final fixing, the breed had been called the Cunningham and the Dunlop. As mentioned earlier, an Ayrshire bull arrived in Dunedin in 1848 with the first party of Scottish pioneer settlers. The first cow arrived the following year. The conformation of the Ayrshire, and especially its udder, is admired, and it took its place in New Zealand dairy herds as an efficient converter of grass to milk.

The Jersey, with its feminine face, lovely moony eyes and fawn colouring, came originally from one of the Channel Islands off the coast of France, although the island was once known as Alderney as was its most famous cattle. Some claim the animal has been a purebred for six centuries and it was certainly known in England as a superior producer of butterfat about 250 years ago. During the nineteenth century, Jerseys were imported by every country with a dairy industry and proved adaptable to a wide range of conditions. In America, it was bred up in size and lost some of its delicate conformation — with straight top lines right through to the rump and deep in the body. It is an extraordinarily productive cow for

During the nineteenth century, Jerseys were imported by every country with a dairy industry and proved adaptable to a wide range of conditions. At the end of the 1950s, seventy-five per cent of New Zealand dairy cows were Jerseys.

its size, and it reigned as the queen of the New Zealand dairy industry as long as farmers were paid solely on the basis of butterfat content.

At the end of the 1950s, seventy-five per cent of New Zealand dairy cows were Jerseys. An interesting contrast exists between the female and the male, which, although smaller than most dairy bulls, is tough and aggressive.

The Holstein-Friesian has supplanted the Jersey, but the Jersey remains a major component of the national dairy herd of approaching five million in 2018, both as a purebred (about fifteen per cent) and as Friesian–Jersey cross (about twenty-five per cent). Another Channel Islands breed, the Guernsey (also eponymous), has been present in New Zealand for some time but has yet to make an impact.

Friesians originally came from the West Friesland region of Holland and Holsteins from Holstein, Germany, 300 kilometres to the north. However, the breeds are now indistinguishable as many countries have used animals from both regions to develop the same black-and-white breed. The first Dutch Friesian cattle were imported by famous Canterbury farmer, John Grigg, in 1884, and this is the reason New Zealand farmers have tended to use the name 'Friesian' even though the breed name here is officially New Zealand Holstein-Friesian.

Since the 1960s, North American Holsteins have been introduced, at first through Canada, and more lately from the United States. The North American black-and-whites — farmed mostly indoors and not under the intensive grazing system used in New Zealand — are huge animals which produce impressive volumes of milk. This introduction made itself felt in New Zealand production over the last three decades of the twentieth century and saw the Holstein-Friesian become the predominant dairy animal. However, in the United States, the breed had an average lactation of only around two years because animals had to be culled young when they did not get pregnant. This fertility problem was generally passed on to herds here and resulted in a decline in the national average lactation. Thus the industry turned its attention to breeding for fertility in a bid to enhance the already established economic value of the Friesian.

Another European dairy breed that is well established in many countries and known for its long lactation life is the Brown Swiss, introduced in the mid-1970s. Another Swiss breed, the Simmental, was introduced into New Zealand as a beef animal in 1970, but it was considered a dairy animal too in Switzerland. At least one Taranaki farmer began milking Simmentals, growing his bull calves as beef.

New Zealand farmers have shown a starkly pragmatic approach to animal breeding, helped by the brilliant recording systems of Livestock Improvement and also by milking-shed technology that increasingly enabled them to monitor the production of each cow at milking time. They have therefore bred and culled according to the productive capacity of their individual animals and not according to its breed or superficial conformation. Friesians came into favour entirely because of their ability to produce high volumes of milk. The number of Friesian–Jersey crosses steadily increased because of individual productive capacity. Early research into desirable traits for once-a-day milking hinted that the Jersey might make something of a comeback because under those conditions it initially did very well.

Dairying also has a long history of using science to its profit in New Zealand and although the level of research — especially what is called blue-sky research — declined from 1986, organisations such as Livestock Improvement doggedly tagged practical problems and worked towards the future. The moves, the dreams, if you like, that the industry was pursuing included robots to do the milking, perhaps in mobile milking sheds that would go to the cows instead of the cows having to go to the shed. Robots were already in action in the closed-in sheds of Europe.

Like most of New Zealand's economic activity, the dairy industry changed dramatically from the 1980s. In 1994, there were fewer than four million dairy cows. Twenty-five years later there were close to seven million. The great area of expansion occurred in Canterbury, which had supported dairy mainly for town supply for most of the century before the 1990s. Then came the boom that brought more than 200,000 hectares into dairy farming in the province supported by irrigation from cheap supplies of underground water.

More and bigger farms were carrying more cows that were more productive than elsewhere after attracting corporate investment to farming's most reliable cash flow. It was factory farming. Expansion occurred also in Southland and Otago where irrigation that defied some of the vagaries of weather helped make the transition to corporate farming. There was even a move into the Mackenzie Country.

A large private farming and processing company, Synlait Milk, was established in Dunsandel, south of Christchurch, in 2000, and processes milk from the burgeoning Canterbury supply. Almost twenty years, later it was establishing plants in other dairying regions.

The downside to this explosion of land use conversion to dairying — the number of cows, the land they cover and the exploitation of water resources — has been an increasing national resentment at what many see as the degradation of the natural environment, especially the nation's waterways, and a failure to curb carbon emissions that contribute to climate change.

Also, the possibility of exotic disease introduction is much greater now that dairy herds are huge and stock movements consequently more difficult to trace, as has been indicated by the $1 billion attempt to eradicate *Mycoplasma bovis* disease, first detected here in 2017, a threat to the beef and dairy industries. An incursion of foot and mouth disease (*Aphthae epizooticae*) would have an even more catastrophic effect on dairy and beef farming and thus on New Zealand's economy.

Nationally, in the rush to the future, the most worrying concerns looming immediately ahead that are hardest to solve are water shortages and the industry's interface with the environment requiring new techniques, new ways of fixing nitrogen in grasses, and increasing costs. And, of course, the vexed question of how to adapt towards global climate change.

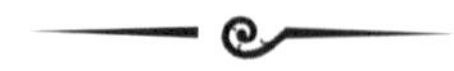

## CHAPTER FOURTEEN

# Scientists and Good Farmers

Until the end of the First World War, farming in New Zealand was controlled by practical men who had little time for the intellectual pursuits associated with scientific research. They observed problems and did their best to solve them in a common-sense way.

Commercial farming on any scale began with running sheep over native tussock in the 1840s and 1850s. As L.G.D Acland put it in *The Early Canterbury Runs*: 'The runholder's practice was simple at first. He put his sheep on the run and tried to keep them there. He marked the lambs when necessary and shore the sheep once a year. … He tried to breed up his flock as fast as he could. Until about 1868 when all the runs were fully stocked there was always a good market for store sheep.'

Many of the squatters were intelligent, well-educated and possessed at least some capital. Acland said that among them were 'baronets, younger sons of good families, soldiers, sailors, parsons, lawyers, tradesmen, shepherds, farmers and … several foreigners'. But no matter how intelligent and experienced were the runholders, their lives were relatively straightforward contests against snowstorms or to contain diseases like 'scab'.

Successful all-round farmers like John Grigg used their experience and repositories of knowledge gained in England to breed up their animals and grow their crops, and they did use records to refine their work. Those carving farms from the bush in the north got their knowledge from shared experiences. But what both sheep and arable and bush farmers did recognise and accept readily were advances in mechanisation that reduced costs and gave relief from the back-breaking labour on the land.

Otherwise, interest in science, or even the toleration of it, was hard to generate among the mass of practical, grafting farmers who believed all problems bowed before hard work and common sense. From the same outlook arose the indifference towards agricultural education at secondary school or tertiary level unless it could be seen to be wholly utilitarian. Towards the end of the nineteenth century, a Hawke's Bay farmer and Member of the Legislative Council, the Hon. Mr Wilson, in condemning the Lincoln School of Agriculture, said he did not agree with 'making farmers through a course in chemistry'.

But in the early years of the twentieth century it became apparent that New Zealand soils were not as naturally fertile as had been assumed and that soil deterioration was resulting in erosion and waning productivity. Between 1840 and 1910, about ten million acres of forest, fern and scrubland was grassed by surface sowing after burning the natural vegetation. Immediate results from the potash deposits were good but the assumption that the pasture would hold on and flourish became observably untrue in many cases and science was the wizard to which afflicted farmers looked.

In 1905 a Mr P. Pattullo, of Otago, said: 'The land does not now smile with harvest when tickled with a hoe as it did twenty years ago. … We find also that our pastures do not hold in grass so well as they used to. … Land is getting dearer, expenses heavier, and it will require more attention, more knowledge, and more trouble on the farmer's part to grow crops successfully than in the early days.'

In response came the first assault on agricultural problems by science, with work mainly by chemists and botanists on soil analysis, the variable requirements for fertilisers and the productivity and longevity of available types of pasture. Rock guano was brought in during the 1860s and 1870s, but the need for regular fertiliser application was underlined by the director of Canterbury Agricultural College at Lincoln, William Ivey, who imported a shipment of superphosphate in 1880. The following year, commercial production of the fertiliser began in Dunedin.

Bernard Aston, a Department of Agriculture chemist, and others, made profiles of New Zealand soils and found great variation in textural properties and in the content of organic matter. Aston was also involved, along with Edward Vaile, in solving, more or less by accident, the animal health problem of 'bush sickness' that stalled the development of large areas of the North Island's central volcanic plateau.

Vaile owned 53,000 acres of the pumice country and managed to keep his stock in good health by using limonite, and iron compound, on Aston's advice. Aston thought the deficiency was iron, but by the 1930s he and Vaile realised that the problem was lack of cobalt, of which there was enough in limonite to make a difference. The sickness was completely countered by the application of cobalt. Trace elements, such as molybdenum and copper, were found to be missing in other regional soils.

By the 1920s, the Wheat Research Institute at Lincoln College had intensified interest in plant breeding and selection. An early inspiration was Dr F. E. Hilgendorf, botanist and zoologist, a versatile teacher and researcher, whose work with cereals brought gifts to New Zealand farming.

Dr Leonard Cockayne, and later Sir Bruce Levy, steadily edged forward with internationally advanced work on the selection and breeding of pasture types and the use of nitrogen-fixing clovers as a valuable, cheap alternative to nitrogenous fertilisers. Levy's mana among farmers ensured that his good advice was widely spread and earnestly taken. It was slow, undermanned and underfinanced work but, in the end, wonderfully effective.

By 1932, Bledisloe could say of grasslands work: 'There is no field of research and experiment more closely affecting New Zealand husbandry and none in which New Zealand scientists and farmers have more skilfully pointed the way to the rest of the Empire.'

What was lacking, though, was an overall plan for the support of industry by science. The New Zealand Institute, a scientific organisation formed in 1868, made some suggestions about the organisation of scientific research. Then in 1925, the government asked the director of England's Department of Scientific and Industrial Research (DSIR), Sir Frank Heath, to visit New Zealand

and advise on how best to organise science to support industry. The government responded to his report with the Scientific and Industrial Research Act of 1926, which set up the New Zealand DSIR and signalled the beginning of government patronage of scientific research that lasted for more than fifty years.

Heath pushed hardest for science to assist the primary industries, starting with dairying as the most urgently in need. Dairy exports the year he visited were worth £16 million and no one could escape the conclusion that in the foreseeable future the economy was going to stand or fall by the export of wool, meat and dairy products, as well as some crops and pipfruit. Heath suggested an Institute of Dairying be set up and used as a model for other organisations to promote the production and processing of grain and grass crops, fruit, sheep and cattle, and meat and wool.

The leaders of the dairy co-operative movement were already aware that in Denmark and the United States laboratories were exploring the potential for milk products. In 1921, the biggest amalgamation of co-operatives in the country had come together as the New Zealand Co-operative Dairy Company and established a laboratory in Hamilton for routine testing directed towards turning out a wide range of standardised, high-quality products.

The company was already manufacturing butter, cheese, casein, dried milk and dried skim milk. Companies in Taranaki had formed a professional association, the Federation of Taranaki Co-operative Dairy Factories, and this group set up a laboratory in Hāwera in 1925 with help from a Department of Agriculture grant of £1000 a year.

The government decided to establish Massey Agricultural College at Palmerston North and directed it to concentrate primarily on dairy issues. Nearby was the Dairy Research Institute which soon affiliated the laboratories in Hamilton and Hāwera. An experimental dairy factory was set up. However, much of the energy and enthusiasm waned when the economic depression stalled capital investment in all industrial activity.

The next big move came with the appointment in 1943 of Dr Percy McMeekan as director of the Animal Research Station at Ruakura. McMeekan was a foundation student at Massey who later graduated with a doctorate from Cambridge University. He was an extrovert who listened to the problems of ordinary farmers and, in reply, spoke their language. His communication and promotional skills were so strong that he became a celebrity scientist and administrator. He turned what was just a government farm into a property on which empirical science could be applied to the basic problems of animal breeding and production.

McMeekan assembled a cast of scientists, many of whom became known in every pastoral farming country in the world — John James working on artificial insemination, Lindsay Wallace on animal nutrition, Wattie Whittlestone and Doug Phillips on the whole process of milking cows, John Hancock and Alan Carter on genetics. The stellar decade of work at Ruakura was 1948–58 when it was acknowledged by some scientists in other countries as the best animal research establishment in the world.

At the same time, the Waikato produced a number of extraordinary farmers. One was Jack Ranstead, who assembled an incomparable private library and from that and his work on his farm developed both theoretical and empirical knowledge of animal breeding and husbandry that ranked him among the top world experts. His father, William Ranstead, had been a friend of the famous English writer and artisan, William Morris, and after a visit here in 1899 returned to England determined to test his Fabian faith by emigrating with a group of like-minded, high-minded people to establish a utopian settlement in New Zealand. Utopias were all the rage at the time.

The party of 200 came out on four ships, in 1900, and later arrivals swelled the group to 2000. This utopia, like almost all of others, collapsed almost as soon as it began so Ranstead eventually settled at Matangi, near Hamilton. Jack Ranstead, one of William's five sons, gained a diploma in dairy farming at Lincoln College and began his intellectual and practical interest in animal production. He became the first president of the New Zealand Animal Production Society during the early years of the Second World War and was an active participant and patron of research.

A neighbour of the Ransteads at Matangi was Dr H.E. Annett, who settled in this country in 1927 after twenty years in the Indian Agricultural Service. For many years Annett conducted fertiliser and rotational grazing experiments on his property at his own expense.

Before the flowering of Ruakura, practical, applied animal research had been submerged by the breed societies which kept a reasonable standard of performance with the most widely used breeds, but relied too much on the often-irrelevant qualities of conformation and trueness to breed type. British judges were frequently brought in to agricultural and pastoral society shows and judged long-established New Zealand adaptations of cattle and sheep breeds according to the conformation standards of British types. With scientific backing, the mystique of conformation was almost totally replaced by the pragmatism of production recording.

Of course, some clever practical farmers had known this for many years. A leader of the National Dairying Association from 1899 to 1921, Mr T.G. Harkness, told the association conference the year he retired:

> *Do you know what I think is the best work I ever did for the industry? When I stood on this platform and read to you a paper on how to improve a dairy herd long before anyone else had taken the matter up.*
>
> *I said, 'Don't waste your strength on feeding a cow that is not profitable, but see, every one of you, that you have a herd that can do 250 to 300lb of fat and then you will be able to meet the difficulties that are facing us in the near future.'*

That year, 1921, the average butterfat per cow was 174 pounds.

And yet by the 1950s, productive performance was still a secondary issue with some. McMeekan and his scientific staff deflated breed society pretensions with Ruakura's unique (at that time) commercial herd size experiments on animal breeding, production and feed conversion, grazing techniques, on the milking process and other parallel issues that revolutionised dairying by 1960.

By the 1970s a Department of Agriculture pamphlet on commercial sheep flock improvement said simply: 'After an eye appraisal for wool or physical fault, the farmer should choose a ram on fertility rating, fleece weight and bodyweight recordings.'

The Ruakura group made the paddock their laboratory. If research projects did not apply directly to the business of livestock farming under conditions prevailing in this country, they did not undertake them. They aimed at achieving the highest production per acre at the lowest cost to the farmer.

Scientists at Ruakura did a wide range of other work involving farm machinery development, and animal reproductive physiology, nutrition and health. One of its most publicised achievements, which was a long time coming, was the discovery of the cause of a disease known as facial eczema. It was caused by a toxic fungus, *Pithomyces chartarum*, that grows on pasture following rain and high humidity in the late summer and autumn. First noted in 1882, it became a nagging problem for farmers, especially in the high-production regions where warm rain in autumn and summer kept the grass growing.

Finding the cause became a matter of serious national concern in 1938 when millions of sheep and cattle were hit by the fungus, which produces a toxin that attacks the bile ducts and liver and causes the visible symptom of a sunburn effect and scabs on the animal's face. Scientists worked on the problem assiduously as pressure mounted to find the cause. But it eluded them for twenty years, mainly because they had dismissed too early the likelihood that the disease was fungal in origin.

One day, a scientist, Campbell Percival, noticed a black dust on the blades of a mower, collected some, identified it as the spores of a fungus, cultured it and fed it to guinea pigs. The search was over and a preventive was on the way.

The eminence of men like Levy and McMeekan effected a strong change in attitudes towards agricultural education after the Second World War. At secondary level, the agriculture courses had long been regarded as a refuge for dullards. At tertiary level, educationists seemed to lack the will to move and farmers the drive to push them.

Although the Lincoln School of Agriculture was the first in the southern hemisphere when it opened in 1880, it soon lapsed into an educational coma, and a report in 1925 said: 'The present position makes our task (of planning the future of agricultural education) especially difficult; Canterbury, a Special School of Agriculture, without a professor; Victoria College, with a professor and ten students but no School; Auckland, with a professor and neither students nor School.'

After McMeekan left Ruakura in 1962, many Lincoln and Massey graduates continued to make great progress on animal production at Ruakura and a number of other government research institutions around the country. They helped keep New Zealand among the top agricultural countries in the world.

Milk always demanded more care and scientific knowledge than any other bulk primary product as it journeyed from the milking shed, through factory processing, to international markets.

Wool was easier: you clipped it and packed it. Meat required easily achievable expert butchering and temperature-controlled storage. The quality of these products depended mainly on experienced husbandry. Dairy farmers needed similar husbandry skills, but milk is a fragile and perishable product, requiring careful attention on a daily basis by thousands of small farmers and processing workers.

For more than twenty years after butter and cheese became export commodities, the industry struggled to achieve a consistent level of quality. Inadequate facilities and underskilled people working on many of the hundreds of undercapitalised farms and factories meant assistance and regulation by central government and industry organisations was pivotal to building an export trade.

Twenty-five years after refrigerated shipping opened up international markets, the *New Zealand Official Year-Book* reported: 'The government still spends large sums of money in teaching the art of butter and cheese making, the proper method of grading, packing and shipping.'

Gradually, dairy advisers at the farm and factory level in New Zealand raised the level of equipment and expertise and inspectors in Britain reported on the quality of product on arrival.

The 1907 *Year-Book* also said the government recognised the importance of breeding 'better strains of cattle for the production of milk of superior quality as well as quantity' and thus had imported some first-class sires for dairy farmers not in a position to secure the best bulls. 'The service of these bulls has been fixed at a nominal scale and is only available for selected animals.' So the government was involved in breeding up the national herd at least two years before the first testing of factory suppliers (to the Dalefield Dairy Company, near Carterton). Eight hundred and fifteen Dalefield cows were tested for butterfat yield in 1909.

The desire to raise production nationally by identifying high-production cows spread. In 1923, a group herd testing scheme with a staff testing officer began in the Waikato, covering 7500 cows. The government and industry realised the value of group schemes, became deeply involved and by the end of the 1920s the New Zealand Dairy Produce Control Board, progenitor of the New Zealand Dairy Board, reported that increased butterfat production per cow was testimony to the value of the herd-testing movement: between 1922 and 1930 per cow production grew twenty-nine per cent, from 175 pounds to 225 pounds.

By 1929, the twenty-seven herd-testing associations were amalgamated into six, each with a consulting officer to act as an advisory link between the associations and the Department of Agriculture. The scheme was funded jointly by the government and the Dairy Board.

The government's support helped the movement gain traction. In 1935 E.J. Fawcett, Department of Agriculture farm economist (later Director-General of Agriculture), wrote that the systematic testing of dairy cattle on yield had become an integral part of management on many farms and rendered great service to the dairy industry by providing a measure by which any farmer was able to study their farm and herd management. The element of competition introduced through the press and agricultural publications had spurred farmers on to greater effort and improved management. 'Thus testing has become a factor of incalculable value in the life of the industry.'

Following more than a decade of research into insemination techniques by the Department of Agriculture's Dr John James, the first commercial artificial breeding service began in the Waikato and Taranaki in 1949 in a bid to grade up the national herd. A year later the service was taken over by the dairy board. Artificial breeding proved so effective in herd improvement, especially as techniques improved dramatically, it expanded rapidly.

By the 1971 season, 716,000 cows, thirty-two per cent of the national milking herd, were tested and 961,000, forty-three per cent, were artificially bred through the service. Consolidation, centralisation and expansion of the scheme continued through

2001 when the amalgamation of dairy companies led to control over testing and artificial breeding by the farmer-owned, Hamilton-based Livestock Improvement.

By the early years of the twenty-first century around eighty per cent of cows were being herd tested and artificially inseminated, and around ninety-nine per cent of all dairy replacements were identified at birth and details recorded on Livestock Improvement's database.

The progressive development of dairy farming through to the end of the twentieth century depended on: scientifically improved pasture; superphosphate topdressing; herd testing on a national basis to elevate the performance of sires and individual cows through artificial insemination; carefully monitored rotational grazing to gain maximum economic value from pasture; and heavy investment in scientific research on the manufacturing side. The trend was always towards larger herds feeding on grass for as long as possible through the year on bigger properties, with hay for winter supplementary feed and silage for use as the season faded.

As strong-wool sheep breeds, notably the Romney Marsh, became more important to New Zealand farmers, they were also controlled by breed societies that looked back to their origins in England to ensure the breeds remained vigorous, 'pure' and true to type. The system worked like this: stud breeders registered their ewes and rams with the appropriate breed society or association, providing officially recorded pedigrees.

In front of each flock book was a long and detailed description of what the ideal animal of that particular breed should look like — its conformation. These breeders either sold rams direct to commercial farmer or on to middlemen, if you like, who multiplied the animals and sold them on to commercial farmers.

Commercial farmers were not allowed to sell rams for breeding. This mystique of conformity and purity was reinforced by the purchase from time to time by stud breeders of expensive, prestigious sires from the breed societies in Britain to renew the 'pure' genes; and British stud breeders were often brought to New Zealand to judge local stock at Agricultural and Pastoral shows to make sure breed standards were true and consistent. The system was such a brilliant exercise in marketing by the British breed societies it became international, especially among those countries within the empire. In this way, the United Kingdom became the lucrative stud farm for the world.

The concept even spread to humans. The immigrant population in New Zealand was even more narrowly homogeneous than the livestock. Almost all the early settlers came from England, Wales, Scotland or Northern Ireland. They brought with them a belief in the moral power of British Protestant righteousness that was all the better, they felt, for regular cultural infusions from the Mother Country.

As late as the 1930s, Ewen Alison, businessman, former Mayor of Devonport and then Takapuna, MP and Member of the Legislative Council, could write:

> *New Zealand has a heavy atmosphere. It is about fifteen pounds to the square inch. This heaviness of atmosphere tends to make people sluggish and slow moving. This applied more particularly to Auckland than it does to Wellington and the South Island...*
>
> *Our young people of the second, third and fourth generation are not so virile and robust as their parents and grandparents. We are a retrograde nation physically, and it is necessary to have infusions of blood from the Homeland to keep up our excellence. Our horses, our cattle, our sheep and our dogs require the same restoration.*

While very few people accepted the degradation of human colonial genes, the concept of renewing pure British livestock strains persisted into the second half of the twentieth century. However, practical farmers gradually became aware that the economic performance of their animals was not always consistent with their so-called purity. McMeekan and his associates had enraged breed societies by telling farmers that they should breed their Jersey cows basically for butterfat production not for their comely conformity. Herd testing of dairy cows was, by then, well under way and the evidence was compelling.

In the 1960s, Professor A.L. Rae of Massey Agricultural College helped enlightened farmers to set up the National Flock Recording Scheme (NFRS) which offered to sheep stud breeders and anyone else interested the chance to record for the first time in an organised way performance data such as lambing percentages, live weights and fleece weights. Coopworth (Border Leicester x Romney) and Perendales (Cheviot x Romney), new breeds fixed during the 1950s, were taking their place in many commercial flocks and the Coopworth Society made it mandatory to be on the NFRS.

But the impetus came mainly from commercial farmers who tagged and recorded the performance of their sheep for processing by NFRS to identify the top ewe lambs for future replacements. One of the main considerations for the sheep farmer was to identify ewes with good lambing percentages and easy births to fit in with the easy-care New Zealand farming style.

The stud breeders' income was hit as farmers were surprised to find how high was the quality of their top ewes and rams and began breeding up within their own flocks, and even selling rams to other farmers who wanted sheep bred for productive lives in the economic reality of the farm rather than in small studs.

The next move was Group Breeding Schemes (GBS) in which some groups of commercial breeders identified their top ewes (usually two-tooths) and contributed them to a central flock, usually of a group members' farm, for ease of recording. Rams bred in the nucleus then went back to be used on contributors' farms at a ratio of one ram for four contributed two-tooths. The nucleus farmer was paid a management fee for the extra work of recording.

As this revolution developed, stud breeders missed out on sales, breed associations lost registration fees and stock and station agencies got no commission. Lobbyists sought to frustrate the new developments and to curb the support from government-paid scientists, but breeding for economic traits became an unstoppable trend and was essential to survival once farming subsidies were pulled away in the 1980s.

The whole livestock industry was transformed by science. In the end, so were the breed societies which began to provide full performance-recording services and advice on all aspects of breeding.

## Chapter Fifteen

# Another War and the Aftermath

New Zealand paid dearly in blood during the Second World War with an estimated 75,000 people joining the fight against Germany in North Africa and Europe, and against Japan in the Central and North Pacific. About 36,000 died, were wounded or imprisoned during the nearly six years' struggle. In July 1942, mobilisation peaked at 154,549, including 8500 in women's services and essential war work, but excluding the Home Guard and Auxiliary Patrol Services. It was a colossal effort on the part of a small country. But, while many cities in Europe and Asia were razed by aerial bombing and artillery fire, New Zealand's infrastructure was not damaged — even though it was neglected in most of the years through the stuttering recessions of the 1920s, the big Depression of the 1930s and the war.

Vigorous farming continued during the war years and production remained fairly steady despite shortages of labour and fertiliser and lack of investment. Producing wool for uniforms and blankets, and traditional meat and dairy products to feed embattled Britain, our own service personnel overseas and American troops in the Pacific, was seen as part of the war effort. A scheme called 'Manpower' absolved men from conscription to the armed services if they were working in essential industries such as farming, although many farmers enlisted and left family members, including fathers, mothers and wives, to work the land.

Manpower was also used to direct workers not liable for conscription to industries deemed essential. Land Girls were recruited to help keep farm production up. Although resistance to an organised force of Land Girls was expressed at first by some politicians and farmers, nearly 3000 members of the Women's Land Service (WLS) worked on farms after the service was formed by government regulation at the end of 1941, a few months before full Manpower controls were put in place. Although service was voluntary at first, members of the WLS could be compulsorily assigned to farms by Manpower officers from the beginning of 1944. Just over half of the Land Girls worked on non-family farms.

The Labour Government had formed a Marketing Department in 1937 which controlled the export of dairy produce. This department and other centralised marketing controls already in place enabled the government to mobilise primary products as soon as the war started and export them to Britain at set prices. These guaranteed prices relieved farmers of concerns about revenue, even though the meat and dairy prices were generally below world levels until a few years after the war. So New Zealand farmers' contribution to the war was considerable.

During the war, the Māori exodus from the country to the city began. It gained momentum in the immediate post-war era. For decades many had lived at subsistence level in rural areas. Despite land assistance programmes nurtured by Sir Āpirana Ngata in the 1920s and 1930s, the limited land base and lack of development capital following the confiscations of the nineteenth century and continued purchases through the Maori Land Court meant they were never able to regenerate the drive and productivity of the years before the wars of the 1860s and 1870s. The Māori birth rate at the time was three times higher than Pākehā. By the end of the war, 17,000 Māori lived in urban areas, more than twice as many as ten years previously, and during the twenty years to 1971, urban Māori numbers grew from 20,000 to more than 130,000. Many of them worked in processing plants for pastoral industries, especially as butchers in freezing works. As automated plants took over processing and replaced the big exports works, thousands of them lost their jobs.

A wartime bulk-purchasing agreement between Britain and New Zealand for primary produce continued until 1954, so the incentive to seek out new markets remained low. Cheap synthetic fibres developed during the war came on to the market and raised some concerns about the long-term viability of wool, but they disappeared, temporarily, when the Korean War hefted demand to such a degree the New Zealand Government temporarily froze some of sheep farmers' incomes to prevent inflation running amok.

They were good times, the 1950s and 1960s, for traditional primary commodities and thus for the economy. Revenue was bolstered for the first time by forest products, including pulp and paper from the mill at Kawerau, in the Bay of Plenty, which came on stream in the 1950s. New Zealand was then one of the richest countries in the world per capita. In 1969 the National Government appointed a Royal Commission into the future of social security and the welfare state with the intention of extending it in a rational way. The family benefit was doubled. The Royal Commission's report was disappointingly flatulent, but it contained valuable statistics. Accident compensation, a no-fault insurance scheme for workers, was introduced and then extended over the following few years to cover all citizens.

In the immediate aftermath of the war, most commentators considered feeding a hungry world with a resurgent population augured riches for the future of food-producing countries. Farm productivity had increased at varying rates since the early 1920s. The number of people working on the land gradually diminished here as well as in other developed countries, while production volumes climbed as a result of increasing professional skills and technological advances. About one-eighth of the New Zealand labour force was employed in agricultural production by the 1970s, down from about a third at the beginning of the war.

By the end of the 1960s, most New Zealanders could see no end to the golden years, but by then the future was blurred for careful observers by the slow growth of the British economy on which prosperity depended, by increasing protection for British farmers, by the near certainty that Britain would join the European Common Market, and by the abysmal lack of marketing information available in New Zealand with which to diversify products and markets.

The country's economic outlook changed abruptly in the early 1970s. Even before Britain finally clinched the 1973 deal with the EEC, international oil prices surged and knocked developed economies from their relative equilibrium. Wool prices slumped again, this time beginning a decline from which they have never fully recovered. The terms of trade fell by a staggering forty-two per cent by the time of the 1975 election and social dissension erupted as unemployment rose and damaged the national confidence.

The belief gained momentum and finally prevailed that growth would come only through manufacturing and financial and tourism services. Although New Zealand had carefully negotiated quotas with Britain for meat and dairy products for a fairly long period, other developed countries — and economies like Japan's, growing phoenix-like from the ashes of war — were unambiguously refusing unencumbered entry to agricultural products. The Norman Kirk/ Bill Rowling Labour administration borrowed and hoped. The Rob Muldoon National Government from 1975 failed to accept the economic realities of world changes and floundered for a decade, mostly trying political nostrums. Having condemned borrowing by Labour, Muldoon borrowed even more — and waited Micawber-like for something to turn up.

Some progress was made. A few financial controls were eased and the New Zealand-Australia Free Trade Agreement (NAFTA) stimulated the export of manufactured goods to Australia. NAFTA evolved into the free trade of Closer Economic Relations (CER). The diversification of both products and markets began, but slowly.

The economy was in deepening trouble by the time the 1980s arrived. As conditions deteriorated, wage and price controls were adopted, vainly, to damp down inflation. The public was restless as Prime Minister and Minister of Finance Muldoon tried in his cajoling, often bullying, way to conduct the economy like an orchestra playing his own arrangements in a bid to get the country back into tune. To no avail. It was all temporising and when change came abruptly in the mid-1980s, farmers were hit first and hardest.

Perhaps livestock incentive schemes and supplementary minimum prices (SMPs) for meat producers best illustrate how far the business of farming was removed by price supports from the chastening influence of prevailing international prices. In 1978 when inflation was almost out of control and the New Zealand dollar was artificially high against other currencies, experiments with various price supports ended with the government imposing SMPs to insulate farmers against rapidly growing costs. The plan was to give farmers confidence in a future in which the commodities price cycle would once again swing up and balance out losses — as it always had done, sooner or later, in the past. Although the trigger point for SMPs did not immediately arrive, the system became a bulwark against despair. But the trouble was that stock numbers climbed artificially as supply was disconnected from demand. Facilities for handling the kill and distribution systems expanded and new markets could not be found fast enough, not at satisfactory prices anyway.

Price supports allowed a traditional attitude to become a delusion. Bulk exporting, with prices often agreed between British and New Zealand governments, had been the standard for so long the primary industries and the government found it hard to get a grip on the concept of producing food for specific customers.

As late as 1983, the Meat Producers' Board commissioned a business consultancy, Glendinning and Associates, to report on the distribution and marketing of lamb in Britain. The result was damning. New Zealand lamb was 'the worst presented and least attractive commodity available in the supermarket'. Without slick, modern packaging and presentation the chance of getting top returns for meat from discerning consumers was negligible.

Meanwhile, the national flock eventually exceeded seventy million. SMPs for meat were more than twenty per cent over prevailing prices and by the end of the 1983–84 season had cost $595 million, about $450 million of that to top up lamb prices. Marketers bustled around the world looking for new deals to take produce gradually being shut out of Britain plus the new meat mountain created by subsidies. A lump-sum support that replaced SMPs in the following season cost a further $130 million. One estimate is that in their biggest year, 1982–83, the whole range of farm subsidies cost well over $1 billion.

Worldwide, those who work the land tend to have a more conservative cast of mind than urban people and, politically, in a sense relative to the national political centre, New Zealand farmers were no different. They could understand and accept mechanisms of production and price that had been implemented at various times over the years to level out returns between the peaks and troughs of supply and demand; but as governments and bureaucrats, including their own, dithered in the face of impending changes to New Zealand's international trade in the 1960s and, particularly the 1970s, many farmers became uneasy at the growing level and type of intervention. They went along with it because subsidies were seductive, but with increasing discomfort.

What had become clear to most economists and many farmers was that old devices were being employed to fix new problems. Sheep numbers were so inflated by the 1984–85 season that when the end of the supports was in sight the lamb kill was a record thirty-nine million and the sheep kill nearly eleven million. Two years later when the rug of SMPs had been yanked away, the lamb kill was down by six million. Marketing teams had fanned out around the world with faint hope that they could sell this mountain of meat. But it would be nearly ten years before *The Economist* could write: 'They may be out-numbered 16 to one by sheep, but New Zealanders do not follow the flock. In economic policy they lead it.'

The Labour Government that gained power at the end of 1984 devalued the currency and declared it would end subsidisation schemes by the end of the 1985–86 season from when primary producers would have to confront world markets with a strictly commercial stance. The intention was to deregulate the country's economy as quickly as possible and because that economy was anchored in agriculture, farmers took the first and biggest hit.

This new government was ideologically in favour of free markets and while something had to be done to more sensibly align the country's economy with the reality of world prices, the suddenness was draconian. Cabinet's justification was that, as Macbeth said: 'If it were done when 'tis done, then 'twere well it were done quickly.' The government was aware that fulminating public and political opposition would make the policy hard to implement with such dispatch if it was opened for discussion.

It was a revolution and like most revolutions went too far too fast with the needless nationalisation of government infrastructural companies. With a unicameral Parliament, Cabinet could move with rapid legislation, much of which was ill thought out.

The policy bore down harshly on farmers, especially those who had recently taken on debt in the tradition of young smallholders who spend a lifetime working towards high equity. They paid an exorbitant price for the shilly-shallying of governments over twenty years. The new administration understood the electoral power of rural voters would not be enough to unseat it. Only when urban New Zealanders felt the shock of ineptly introduced deregulation in late 1987 did the government's support fall away.

Heavily indebted farmers who had had no hint before 1984 that this would happen were immediately in trouble and, for many, mortgagee sales were the only recourse. One result was a jump in farmer suicides.

The government stripped away any services previously provided by the Department of Agriculture that could be construed as financially supporting farmers. In 1987, the department's advisory services were withdrawn and advice was privatised. Inspection fees and adverse events financial support were withdrawn; and funding was pulled even from government-funded research, which had helped elevate New Zealand pastoral industries to the top of world performance.

All this support was taken away at a time when farmers' incomes were falling after an initial boost brought about by the dramatic currency devaluation. So the 1986 season brought bad weather, the consequences of a twenty-eight per cent devaluation of the dollar, high interest rates and a future obscured by absolutely no further protection for farmer income. Heavily indebted farmers who had had no hint before 1984 that this would happen were immediately in trouble and, for many, mortgagee sales were the only recourse. One result was a jump in farmer suicides.

Thousands marched on Parliament in May 1986, burning effigies of leaders. Topdressing planes buzzed the Beehive with flour bombs, demonstrating that not only farmers but all the ancillary industries were suffering. The effects juddered through many industries and damaged small towns already suffering from the improved roads and cars that could get people quickly and frequently to nearby cities.

Domestic deregulation freed up the long-protected egg and wheat industries, and some competition was brought into town milk supply when a consumer price subsidy was withdrawn along with zoning restrictions on supply and distribution.

In the long run, free market agriculture was in New Zealand's interests, even though it still bumped into barriers erected against primary products in most other developed countries. But the peremptory nature of the changes seriously damaged some sectors of the community, farmers and urban workers alike. To contradict Macbeth: 'If it were to be done at all, 'twere best to be done well but not too quickly.'

But to see this period in its social and political context it should be remembered that conventional wisdom considered traditional pastoral farming would not continue to be the foundation of the economy it had been for more than a century. Having to find its true level of value, farming should be allowed gradually to slump to a vastly lesser role if New Zealand was to survive as a First World country — or so thought most economists and politicians at that time. One Cabinet Minister described agriculture as a 'sunset industry'.

This attitude and the continuing urbanisation of the country, with city voters growing in numerical strength, had ended the power of the farming lobby which had been pre-eminent since just before the First World War. The President of Federated Farmers could no longer phone the Prime Minister and expect to be pencilled in for that day or the next. He was now on the B list.

However, many young farmers sensed a kind of freedom in their dissociation from government, a freedom from what had become an accustomed need to lobby. Bulk marketing to a single major destination required some government control and little true initiative from producers beyond negotiating with politicians over the shape of that control. Now they had a need to respond directly to what the people in other countries wanted from them.

The result was the blooming of a long-latent entrepreneurial spirit. Those who dealt with the big traditional products — like butter, cheese and other derivatives of milk, and meat and wool

— examined and many changed the way they went about their business. Others tried deer farming, the first in the world to do so, goat farming for milk and meat, even ferret farming for fur, ostriches for meat, and other enterprises, all seeking a niche somewhere in either the domestic or a foreign market. Some were successful, some failed. Fruit, berryfruit, flower and vegetable growers who may have been preoccupied by local consumption raised their vision and built up export volumes, using air freight effectively, and slipping into seasonal supply holes mainly in Pacific Rim countries. They looked around for new products.

The glamour stories of diversification were those of kiwifruit, grapes and deer. Early kiwifruit growers made millions and set everyone to fossicking for big ideas, big opportunities. Grapevines now pattern thousands of hectares with their long green grids in many parts of the country but most famously in Marlborough, Central Otago, and the east coast of the North Island.

These grapes in classical varieties feed a wine industry that, over twenty years, has rampantly accompanied the extraordinary culinary revolution. Deer farmers have had difficult years, but they had added another dimension to pastoral farming. More recently diversity in the sector has been fruitful with the gradual expansion of farming goats and sheep for milk products.

No doubt was expressed that the New Zealand economy needed a broader base but commentators clutched at straws, insisting that farming could not save New Zealand and attention had to be focused on information technology, financial services, niche manufacturing and tourism if it were to remain a First World country.

Well, they were right about tourism.

## Chapter Sixteen

# The End of the 'Home Boats'

In the early 1970s, when Britain's post-war flirtation with Europe was finally consummated, New Zealanders, including farmers, were gripped by a kind of delayed disenchantment. Suddenly they grasped that Britain — for whom they considered they had sacrificed so much during two world wars — had cast off the mooring lines of its offshore farm after ninety years.

What then gradually became apparent was that the burgeoning post-war economy of Japan would need to become one of the alternative markets for New Zealand. This was anathema to the generation that had not long before demonised the Japanese as an unforgivable, inhumane enemy.

Faced with the need to sell their pastoral products to discerning consumers in diverse markets for the first time, they realised they had absolutely no skill or experience in marketing. Domestically business had generally been centrally directed, and international trade had been a matter of government and producer representatives negotiating deals with British distributors.

Prime Minister Keith Holyoake had accepted Britain's retreat behind the restrictive quota and tariff walls of Europe as inevitable for most of a decade and reacted in a way that was quintessentially his. With a leadership style that seems retrospectively so attractive in the context of more recent behaviour, he sought consensus by setting up a National Development Conference.

Industrial sector committees met and, backed by the efficient and diligent senior public servants of the time, trawled for essential data from their sectors and made recommendations to the conference, which was held in 1969 when the country was having its first serious balance of payments problem in more than ten years.

The information presented to the National Development Conference was substantial, relevant and valuable and many of the recommendations provided a base for action that was not taken for years because of change of government in 1972 and an economy rocked by oil shocks. For example, the principle of adding value to primary products in this country instead of bulk exporting without further processing was reinvented by many marketers every few years after that 1969 conference. Indeed, 'value-added' remained something of an economic catchword for more than thirty years.

The problem was that most countries, including Japan, had barriers against importing processed goods, preferring rather to bolster their own economies with the processing stage. For example, it was easier to export logs than timber products.

The remedies taken by farmers, by a generation of marketers and, erratically, by governments were impressive and saved New Zealand from economic disaster.

Farmers and their governments had known for a long time of two principal dangers perennially facing New Zealand: first, the extreme reliance on the British market for produce on which the very economic existence of the New Zealand depended; second, bulk supply of produce to distributors in London for them to do the consumer marketing on the country's behalf.

From the beginning of commodity exporting in 1882, British interests took a large measure of control over meat processing plants, shipping services and the marketing of meat and dairy produce. In 1901, Trade Commissioner Graham Gow, appointed by the new Department of Industries and Commerce, travelled widely within the country to acquire the information and 'samples' for selling missions he was to undertake to South Africa, India, China, Japan and elsewhere.

The department aimed to expand overseas markets for New Zealand meat and dairy produce. Documents suggest both farming organisations and governments found the freezing companies, at least in the North Island, were resisting attempts to diversify markets for meat and to branch out from the export of bulk carcases to a single outlet, Smithfield in London.

The department reported a successful export sample of New Zealand lamb to New York and claimed a great mistake was being made by freezing companies in confining meat exports to a bulk trade to London, to 'secure a substantial profit for themselves', when American, Spanish and Scandinavian markets were inviting business. Trade Commissioner Gow, by 1902 in Britain, insisted unexploited markets existed in many other parts of Britain where companies were ready to take wholesale quantities of lamb. The department also insisted the freezing companies in the North Island had adopted an unfair fixed-price procedure for buying of fat stock from farmers.

The report criticised the 'Home dealers' at Smithfield for opposing branding of the New Zealand product because they would be robbed of the immense profits that could be made out of the meat by selling it as 'home-grown' (and thus not frozen). Branding was a problem that continued over a number of years in more than one market. Australian wholesale purchasers of New Zealand butter in 1902 were reported to be removing the original brands and substituting others.

Trade Commissioner Gow reported visits to Aberdeen, Inverness and Dundee where he found limited supplies for sale of 'highly regarded' New Zealand butter, cheese and lamb. He thought the best way to expand trade to Scotland was through Manchester, adding: 'I am of the opinion that steps should be taken at once to make vigorous and sustained efforts to open a market in the Manchester district and the North of England generally.'

He even suggested, prescient by half a century, that 'a trade in wood-pulp might be opened up'. However, the following year, he reported that after a 'careful investigation as to prices, etc.,' he considered it would not be wise to start a wood-pulp industry because the manufacture required 'very expensive machinery' and the country could not compete with Norway and Canada.

Gow had all the overdrive of a salesperson. Part of his brief was to tout for tourism, and he wrote that in the North of England 'several gentlemen have announced to me their intention of proceeding to New Zealand at an early date for a holiday tour'.

He next moved to South Africa from where he sent a huge report on the possibilities of setting up a shipping service between the two countries to foster a frozen meat trade that he was sure would develop into a mainstream business.

Trade Commissioner Gow had the overdrive of a salesperson. Part of his brief was to tout for tourism, and he wrote that in the North of England 'several gentlemen have announced to me their intention of proceeding to New Zealand at an early date for a holiday tour'.

The following year, the indefatigable Gow visited Japan, China, India, Ceylon and Singapore and was enthusiastic for trade prospects with Japan which already had shipping services to Australian ports and, he thought, could easily be persuaded to continue on to New Zealand. Little progress was made, but the government persisted with efforts to diversify, despite tariffs. In 1914, the Department of Agriculture, Industries and Commerce reported butter exports to Vancouver of 60,000 hundredweight in the previous year after a period of 'surprising' growth.

For the first twenty-five or thirty years of refrigerated exports, the major concern was getting consistency of product quality from the many freezing works and dairy factories that sprang up around the country. To this end the government of the day brought out experts from Britain, installed meat and dairy inspectors at the factories and in London where distributors had most control. But fraudulent abuse of brands continued for many years. It wasn't until twenty years after the refrigerated trade started that New Zealand began thinking about cold storage in London to manipulate the timing of when produce went on to the market.

The danger of economic dependence was not unknown during the first third of the twentieth century, but no one was seriously attending to it. In the period immediately before the First World War, the limitations of the British market became apparent. A serious confrontation between farmers and urban workers in 1912 and 1913 was not just ideological as William Massey's government insisted. Unemployment was high and growing, and workers' wages had slipped behind the incomes of farmers and the self-employed.

The war and the boom immediately afterwards changed that. But a series of recessions during the 1920s prompted some observers to reconsider the continuing lassitude of New Zealand's approach to marketing. The recession in 1921 was so sharp it was a factor in forcing rehabilitated soldiers off their recently acquired farms.

Most extraordinary, given all this, was the ignorance by New Zealand of the British market and its consumers. D.O. Williams, a lecturer in agricultural economics at Massey Agricultural College, even fifteen years later, in 1936 wrote:

> *A definite obscurity prevails as to the particular requirements of our customers. ... At present we labour in much ignorance and do little to mend our ignorance. But if we intend seriously to find some escape from an intractable quantity position, we must convert ourselves to a religion of quality and of market specialisation, and this calls for a knowledge of our customer's requirements far more detailed and accurate than anything we possess at present.*

The problem associated with the absence of hard-headed marketing had begun with the trust the colonials had for the people at Home. This may not have been quite as exaggerated under the Liberal administrations (1890–1911) but Prime Minister William Massey (1912–25), who presided over a government of farmers, was an English-born Anglophile who went to Mother England as often as he could while in office. He was a very able politician but abject in his attitude towards Britain.

However, after ninety years of a laissez-faire approach to agricultural trade, Britain negotiated new five-year agreements at an empire conference in Ottawa in 1932. At the time, Britain was taking all of New Zealand's cheese, ninety-nine per cent of its butter, lamb and mutton and ninety-eight per cent of its beef. Some primary products that had established growing markets in the United States and Canada had suddenly been shut out by prohibitive tariffs a few years before. This almost total dependence on the British market engendered a measure of panic among New Zealand legislators, already trying to cope with internationally depressed prices.

One New Zealand commentator said it seemed likely that the Ottawa conference, 'or perhaps we should say the Depression, will be long regarded as a turning-point in imperial relationships'.

The five-year agreement provided for the United Kingdom to impose a duty on dairy products, albeit with preferential margins for the Dominions; to bring products under a general scheme of quantitative control; and New Zealand and Australian mutton and lamb shipments were to be kept static, effectively a quota. The Dominions on their part agreed to maintain preferential margins in favour of British products.

In 1933, the British signed three-year trade agreements with Denmark, Sweden and Argentina which gave British goods access to these countries in return for entry of certain farm produce. Because of the impact on empire preferences, the agreements were called 'The three black pacts' in New Zealand.

A year later, the British Government offered domestic subsidies for domestic milk used in cheese production; and, in 1935, it announced that its long-term meat policy was to protect its domestic industry with 'a drastic reduction of imports of meat … from all sources'. A levy would be imposed on all meat imports, with preferences for the Dominions, the money to be used to help British producers.

The New Zealand Government was so alarmed by this it replied that such a levy would severely penalise its mutton and lamb producers and would involve the country in serious economic loss. The reply noted that the economic development of New Zealand depended on export returns for primary products which, in turn, affected its ability to service the charges on British capital investments — a broad hint to bankers and other investors. The government's retort that the idea of a levy was especially repugnant to meat producers and the people of the Dominion was said to be the first time a New Zealand government had taken an aggressive public stand again the 'Mother Country'.

As the world eased its way out of the Depression, the levies were never introduced and by 1936, reasonably satisfactory arrangements were made between Britain and the Dominions. The war and its aftermath meant Britain's door was opened wide for New Zealand food until it joined Europe.

New Zealand was in the mood for self-examination, even self-flagellation when the Depression ended. Markets outside Britain had diminished and the country remained exposed to every British price crisis.

Agriculture had from the first days of European settlement been critically important to the functioning of the domestic economy. In the 1920s, before the big Depression struck, nearly seventy per cent of the total national value of production was primary products. And these products represented ninety-four per cent of total external trade. The occupation of one quarter of all 'breadwinners' in the country directly involved the production of food, and a government report at the time said: 'The pervasiveness of farming in our national life is reflected in other ways not capable of numerical statement.' It continued:

> *The organisation of New Zealand on a specialised farming basis, producing a limited range of major products to be marketed mainly overseas and dependent on large importations for its secondary requirements and on large borrowings for its capital need is the explanation of many of our most pressing problems. ... Even though the prices of any or all of his products fall catastrophically, as they have, the farmer feels that he has no option but to continue producing. Moreover his external costs, as well as import prices, are much more rigid than his receipts from overseas selling.*

The truth was New Zealand had always had limited resources beyond agriculture, had a small population and, therefore, relied almost totally for economic health on exports of a range of products that the British, almost alone, wanted. At that time, Britain bought ninety-four per cent of the lamb and mutton that was traded internationally, about eighty per cent of the beef, about half the cheese (and the cheddar market was only a small piece of that); and Britain and Germany bought ninety per cent of the butter traded internationally, so where else was there to sell it?

So during the Second World War and in the immediate post-war years New Zealand governments and producer agencies continued the bulk selling arrangements. In 1943, a Dairy Industry

Commission decided 'the opening of new markets to dispose of surplus production can best be undertaken by a single central body acting in the national interest and having due regard to all branches of production'.

The result was contracts with the United Kingdom Ministry of Food that went through until 1954. New Zealand gained the right to sell up to fifteen per cent of butter and cheese to markets outside Britain in 1951, and the Department of Agriculture reported in 1952 that prices in other markets were higher, in fact brought in £1.3 million more than had the produce been sold in Britain. And yet these sales were cut back to five per cent the following year when Britain appealed for more food. Similar bulk-purchase contracts with Britain covered meat exports after the war. The Department of Agriculture reported: 'While the meat shortage continues in the United Kingdom only very small quantities are being shipped elsewhere, but the possibility of exploring other markets, especially the North American market, is receiving consideration.'

In 1964 — midway between the end of bulk purchasing in 1954 and Britain joining Europe in 1973 — a seminar was held at Waikato University on 'Diversification of Primary Industry'. (Some background — the British Government that year subsidised its agriculture with about £360 million in direct farmer subsidies that worked this way: New Zealand lamb had free access at the ruling market price which was twenty-seven pence per pound but British farmers had their return built up to an agreed thirty-nine pence per pound.)

One contributor to the seminar was Dr A.H. Kirton of the Ruakura Agricultural Research Centre. Kirton started by noting the targets committee of the Agricultural Development Conference had found 'that we will have to increase our animal population by four per cent annually in order to maintain [our] export earnings and present standard of living'. He then asked where producers would sell this additional lamb and mutton, and he followed with a long lamentation on the lack of knowledge available about primary industries in general but the meat industry especially. He noted that the application of scientific principles of breeding in sheep and beef cattle was forty years behind the dairy industry.

Kirton said agriculture needed an efficient market research and sales organisation. Better 'factual' information was needed 'on the quantities, quality and suitability of our present meat products even in our main markets. What proportions of the various grades of lamb carcases are sold in the various regions of the UK throughout the year?' he asked. 'At what level of fatness does the British butcher regard a lamb as overfat, and at what level do the Americans regard the lamb as overfat?' He mentioned also the lack of specific knowledge on access to new markets.

The seminar noted that progress had been made since 1954. Meat exports then earned £59 million, eighty-seven per cent of it from the UK. In the 1963–64 season earnings from lamb, mutton and beef and all by-products were a record £119 million, only fifty-three per cent of it from the UK. But this change was almost entirely due to the sudden opening of the American market to manufacturing beef and some access to Japan. Both these markets had variable restrictions which could change from year to year.

Those involved in the seminar were either people from the industry or those deeply involved in industry studies at universities; so the vision, Kirton's apart, was generally narrow with emphasis on the difficulties of diversification rather than on marketing needs. The dairy contribution stressed the difficulty of finding volume markets for cheese types other than cheddar and suggested that product diversification depended entirely on adequate prices for butter. At that time, butter earned £56 million, cheese £21 million and all other manufactured dairy products £12 million.

The first British gestures to Europe in the early 1960s were rebuffed, but the inevitability of membership was becoming apparent. New Zealand did negotiate a reasonable phase-out deal with Britain for when it did join, with gradually diminishing quotas. One advantage the country had in these talks that continued through the late 1960s and early 1970s was one Minister of Overseas Trade in Holyoake administrations from 1960 through to 1972. He was Jack Marshall, also Deputy Prime Minister through that whole period. At the end of his career, he became Prime Minister briefly before being dethroned by fellow National Party Minister, Rob Muldoon.

By the 1950s farmers and governments were at last beginning to accept that the future lay with exploring what consumers wanted, especially in North America and Asia, within the unfortunate boundaries of tariffs and quotas. So in the 1965–66 season, well over half the beef and mutton exported went to Japan and the United States, although almost all the lamb continued to go to Britain. With the traditional market in jeopardy because of the EEC, the dairy industry was also becoming more nimble.

An example of how New Zealanders were starting to rise from their long-time marketing slumber was a serious examination in the mid-1960s of the potential for bulk air-freighting export meat, especially to Asia. Food was already being stowed as filler-freight on scheduled passenger jet services but only for luxury markets because of the cost. However, the 'jumbo' Boeing 747 was on the horizon, providing a huge leap in capacity and, concomitantly, lower costs.

The meat industry seriously investigated the possibility of using a fleet of jumbo freighters to fly large consignments of chilled and frozen cuts to an already diversifying market. They even examined the concept of constructing a special airport — Rangitīkei was one site — as a central take-off point for worldwide distribution of produce. It was a brave concept, but the economics did not stack up.

In summary, for near enough to a century, the farming of New Zealand consisted of producing large quantities of a narrow range of low-price commodities to satisfy the notoriously undiscerning palates of the English. It was a gloriously simple business, requiring no entrepreneurial or marketing skills. If business went bad occasionally, it was because the world broke stride in a cyclical slump and everyone suffered together, not because Britain had decided to shop elsewhere (although it had certainly thought about it in the 1930s).

The cheap food policy that had been a basic tenet of British political policy for many years was a major reason why many Britons baulked at European Economic Community membership in the early post-war period; so when the UK–EEC marriage occurred, New Zealand at first saw the move as apocalyptic. She had been wholesaler for an assured retail market, sedate supplier without serious competition, the mistress of the British food business; and she had lived better than the wife, the British farmer, despite a period in the 1930s when attempts were made to impose quotas.

Well, the kept lady was going to have to take to the streets and hustle while uttering airily such incantations as 'diversification' and 'restructuring'. Gradually the lady came to understand that her goodies were attractive to a wider range of people than she had thought; that her favours were more frequently and widely sought than she had expected. She could at last flaunt her wares with confidence and skill.

## Chapter Seventeen

# Say 'Cheese' — But Not Too Soon

The swiftest social revolution in New Zealand cultural history occurred in the second half of the twentieth century when Kiwis, whose traditional cuisine was British based and narrow in its range, suddenly began to embrace international food and wine styles.

But the revolution had a false start in 1962 when a tiny dairy company in Ōpouriao Valley near Whakatāne decided to produce a batch of camembert. A Dutch immigrant on the staff, Peter Assink, was called upon to make the famous French cheese. He did so from a book of instructions and sought the advice of a local Swiss whose knowledge was also 'theoretical'.

Assink had no experience of making cheese in Holland. He arrived in New Zealand in 1949 at the age of twenty-two, married, and worked on dairy farms in the Bay of Plenty before joining the staff of the Opouriao Dairy Company. The company was founded in 1900 after frosts had burnt black the valley's staple maize crop and milking cows seemed to the locals to be a less risky if more arduous alternative to cropping.

No one seems to remember what inspired the dairy company management sixty years later to pioneer a cheese as removed from real New Zealand life as camembert was at that time. That first year, though, Assink and his team got the moisture content wrong and the factory had a large batch of green cheese on its hands — which it buried.

Undaunted, the following year they made not only passable camembert but brie (also French) and gruyère (Swiss) as well. The small company also produced processed cheeses and spreads, using additives such as ham, bacon and cumin seed. The European-style cheeses didn't sell in the Bay of Plenty so an agent in Auckland was employed. Sales were negligible but the agent was encouraged.

Alas, the next year, 1964, Opouriao was merged with the Rangitaiki Plains Dairy Company. Exotic cheeses were abandoned as the factory concentrated on mainstream casein production. Assink lost his job and Ōpouriao Valley returned to the calmer, cheddar-chomping cultural times in which the rest of the country had remained. The Auckland agent was intrigued enough to hire Assink and take him to Auckland with the idea of continuing with camembert. It never happened. Said Assink years later: 'Kiwis at the time didn't know what camembert was. They hardly knew what wine was.'

This national ignorance did not last much longer. The Opouriao Dairy Company's only mistake was to be ahead of its time. New Zealand's great leap forward from bland social conformity to modern, expansive society was stalled for only about another decade. For while farmers and governments were struggling to adjust to the twin trials of Britain joining Europe and the deregulation of the economy, the New Zealand community, rural and urban, was transforming itself.

The symbol of the change was wine. The switch from sherries and ports to table wines came quickly. New Zealand whites were immediately appealing by the late 1970s if underdeveloped, and local reds gradually asserted themselves against imports from Australia where experience in making them was deeper. By the 1990s, New Zealanders were making, drinking and exporting wines such as sauvignon blanc and pinot noir that were as good as or better than any produced in countries with hundreds of years of winemaking experience.

Farming and food have always interacted within any society in a kind of circular way. People ate what their farmers made available locally and farmers produced what people seemed to want. It took a political and social revolution to interrupt that rhythm. New Zealand had, from the beginning of European settlement, reproduced Britain's pastoral farming commodities, grown typical English vegetables in their gardens and adopted British food and drink styles as well as their understated, conformist clothing and mode of living.

This way of life was marooned atavistically in New Zealand for the best part of a century, while Britain was still suffering the social upheavals of the Industrial Revolution that suddenly, painfully, urbanised its people. It may have been somewhat dull in New Zealand, but the comfortable and plain food repetitively served was not confined to this country.

Because Dunedin was the wealthiest city from the earliest days, the Scottish influence was profound and to a large degree it calibrated colonial life. A year after the Indian Mutiny in 1857, Victorian England's pre-eminent arts and social commentator, John Ruskin, tried to understand why Indians with their great artistic tradition could be so barbaric compared with the Scots. He said he had been to the north of Scotland the previous summer and felt 'a peculiar painfulness in its scenery, caused by the non-manifestation of the powers of human art…'. It was the first time in his life he had been in any country possessing no valuable monuments or examples of art.

Whereas, among the models of art

> *none was more admirable than the decorated works of India. … So then you have, in these two great populations, Indian and Highland — in the races of the jungle and of the moor — two national capacities distinctly and accurately opposed. On the one side you have a race rejoicing in art, and eminently and universally endowed with the gift of it; on the other you have a people careless of art, and apparently incapable of it. … And we are thus urged naturally to inquire what is the effect on the moral character, in each nation, of this vast difference in their pursuits and apparent capacities? … We have had our answer. … Out of the peat cottages come faith, courage, self-sacrifice, purity and piety, and whatever else is fruitful in the work of Heaven; out of the ivory palace come treachery, cruelty, cowardice, idolatry, bestiality — whatever else is fruitful in the work of Hell.*

Although this tirade was prompted by British ignorance and arrogance, it did give a contemporary picture of the ascetic, moralistic Scots who had settled Otago ten years before.

New Zealanders had from the beginning been strikingly progressive, confident, inventive and hands-on when it came to the practical business of making things work more efficiently on the farm; but visitor after visitor had noted their extreme social conservatism, bred by extreme isolation. Only the wealthy could afford to travel to what they called 'Home' once they had arrived in New Zealand, except for servicemen who saw the narrowed vision of only death and turmoil on their travels.

The breeze of change, faint at first, came in the late 1960s, after the first wave of young Kiwis returned from their OE (overseas experience) in Britain, and after television had ushered in images of the wider world and jet aircraft began to bring in tourists. The country was ten years behind the same cultural revolution in Australia where it had been enhanced immediately after the war by shiploads of Italian and Greek migrants.

New Zealand culinary standards were plain and had moved little from the British middle-class styles of a century. In 1953, English writer Eric Linklater wrote of a visit to New Zealand in his *A Year of Space: A Chapter in Autobiography*:

> *But New Zealanders, like the Scots, think that baking is the better part of cookery, and spend their ingenuity, exhaust their interest, on cakes and pastries and ebullient, vast cream sponges. Soup is neglected, meat mishandled. I have seen their admirable mutton brought upon the table in such miserable shape that the hogget — so they call a sheep of uncertain age — appeared to have been killed by a bomb, and the fragments of its carcase incinerated in the resultant fire.*

Another more vituperative departing broadside at New Zealand's cuisine came from eminent British media person, BBC broadcaster Wynford Vaughan-Thomas, three years after Linklater's book. In *Farewell to New Zealand* he wrote:

*Super-suburbia of the Southern Seas,*
*Nature's — and Reason's — true Antipodes,*
*Hail, dauntless pioneers, intrepid souls,*
*Who cleared the Bush — to make a lawn for bowls…*

*Saved by the Wowsers from the Devil's tricks,*
*Your shops, your pubs, your minds all close at six.*
*The Wharfies' Heaven, the gourmets Purga'try:*
*Ice cream on mutton, swilled around in tea!*
*A Maori fisherman, the legends say,*
*Dredged up New Zealand in a single day.*
*I've seen the catch, and here's my parting crack —*
*It's under-sized; for God's sake throw it back!*

At that time, such withering ripostes hurt New Zealanders who, however, found later they didn't care about the opinions of visiting Brits. They had been there and seen them in their habitat and didn't have much to envy. And, more extraordinary, in the twenty-first century the average Kiwi cuisine at home and in restaurants was better than theirs, and some of the wine much better than France's, although French cuisine still led the way. Both Linklater and Vaughan-Thomas were comparing ordinary fare in New Zealand's relatively small towns with restaurants in London and other large-population cities.

As any Kiwi who was around in 1953 and for a few years afterwards will know, hogget is the most delectable eating stage in the life of a sheep — its age not indeterminate, as Linklater had suggested but approaching one year, just before the two-tooth stage — with the tenderness of a lamb but a deeper flavour. Exquisite, if roasted well with potatoes, pumpkin, parsnips, onions and kūmara around it, a practice pretty well confined to New Zealand. Elsewhere, cooks roasted meat on its own. Linklater must have been in a rare household or hotel restaurant because the one meal New Zealand women, and especially country women, could cook, was roast lamb, beef or pork.

However, anyone old enough to have lived through the 1960s would have to concede that the delicacy of cooking in New Zealand at the time was not high and was at its best when a dish did not need hovering over, could be put inside an oven or a pot and left pretty much to its own devices — roast meat, steak and kidney pie, stews. Cooking was a lowly trade outside the home, and an arduous one inside it. It was almost entirely the work of women.

The only places men were trained to cook were at sea, in the armed forces or in jail. Training in cooking that included foreign dishes did not become generally available and popular until the 1970s. The social difference between a twenty-first century chef and an early twentieth-century cook is the difference between a lord and a peasant, the status of one inflated and of the other unfair. The roast apart, Linklater was right that meat was often mishandled. Here is advice on 'The Art of Grilling' from the *New Zealand Farmer and Stock and Station Journal* of April 1930:

> *It is an art — the cooking of chops and steaks, and even of making toast, under the griller of the gas stove. But it is one that is very quickly mastered, although the young amateur cook always seems diffident about attempting it. There are two things to remember ... the grill itself must be red hot before it is used, and so must the bars of the gridiron. The second is that the food must be watched all the time it is grilling...*
>
> *Grease the bars of the grid thoroughly as soon as the grill is red hot, and put the steak on them. Push it into place quickly and lower the gas slightly if the flames threaten to touch the meat. As soon as the outside has been sealed by the heat — it takes from two to three minutes — turn the gas down by half and leave it like that for five or six minutes.*

Steak tended therefore to be eschewed because it could not be chewed. The range of cooking advice in magazines such as *The Farmer* was narrow, except for cakes, biscuits and puddings. Linklater was right about the predilection of women, again especially country women, for baking rather than cooking. *Aunt Daisy's Ultimate Cookery Book*, published in 1959, contained 1300 recipes in 298 pages, 201 of them dedicated to sauces, puddings, cakes and other sweets and desserts. It was no accident that New Zealand's all-time best-selling cookbook is sponsored by Edmonds, manufacturer of baking powder. The way lives were lived then had a bearing on this. Household meals were simply enough cooked, but baking required higher skills, more measuring, better timing and more supervision.

Restaurants were rare, mainly in posh hotels, expensive because labour costs were high, and patronised by diners only on very special occasions, such as weddings and twenty-first birthdays. New Zealand, by the nature of its widely spread people in suburbs and towns, did not have a restaurant culture. Apart from a lack of discretionary income in most families, people did not have the mobility to move around freely in the evening. City workers went home at five o'clock from work, or six o'clock if they went to the pub, had an early meal, spent the rest of the evening at home and went to bed.

> New Zealand did not have a restaurant culture. City workers went home at five o'clock (six o'clock if they went to the pub), had an early meal, spent the rest of the evening at home and went to bed.

Farmers also tended to eat early, listen to the wireless, perhaps, and go to bed early in order to rise early. When functions were held in the country, they were seldom sit-down dinners but rather ladies-a-plate occasions — meetings, 'socials', concerts or parties. When people called at the weekend, they tended to do so for morning, afternoon tea or supper — another occasion for scones, tall fluffy sponges, large lavish cakes and biscuits, usually embellished with whipped cream.

The revolution that changed the New Zealand way of living began in earnest with a referendum in 1967 that ended fifty years of six o'clock closing of hotel bars, eighteen years after such a move had been resoundingly beaten in a similar referendum. Moves had been made earlier in the 1960s to allow some restaurants to serve alcohol with meals but under restrictive conditions. When it was realised in the years after 1967 that drunkenness and debauchery would not break out wildly in public with extended pub opening hours, the door creakily opened further and allowed gradual extensions of licensing regulations that eventually extended to cafés and corner stores, conditions as loosely applied as anywhere in the world.

Perhaps the first distinct signs of new ingredients came in the forms of olive oil and garlic, both ingredients which New Zealanders had, without trying them, regarded with anathema for cultural reasons. When the *Woman's Weekly* began to advocate the use of these two ingredients in its cooking columns it attracted blistering letters of complaint. Looking back, one can remember the dish that hailed the conversion of a housewife to the new cuisine as the shrimp cocktail, ubiquitous until it became an offending cliché.

But the change was not, as it is often portrayed, the result of a nation emerging from studied ignorance. The cause was the convergence of a number of factors:

- The rapid spread of the home refrigerator-freezer.
- The spread of other appliances, such as washing machines, freeing housewives from much hard labour.
- Improved international and national transport systems, and packaging, reducing dependence on local produce and on seasonality.
- The end of rationing and then the loosening of the import licensing system.
- Foreign tourists arriving in numbers.
- Young Kiwis returning from OE in London.
- An expanding amount of income for discretionary spending in what was then one of the two or three wealthiest countries in the world.

Ridiculing New Zealand cuisine of the pre-1970s by food commentators since the 1990s has become a national confessional, as though Kiwis were expiating some special sin, as though they ate a narrow range of food from sheer ignorance, whereas good reasons explain it in retrospect. Housewives, and especially rural women, cooked in coal-range ovens, sometimes even camp ovens, and needed food that could be left because the technology did not allow for fine-tuning, and because they were so busy with other diurnal chores around the house or on the farm at a time when families were bigger and before household appliances became common.

Pioneer teacher and home science specialist Tui Flower recalls that an American firm claimed from research that a woman washing the heavy, coarse-fibred clothes of the time, using a copper, tub and hand-wringer, was expending the same amount of energy as a woman swimming breaststroke for five miles. The cooking cycle for meals had to take into consideration the keeping qualities of meat, milk and other ingredient of meals before refrigeration became widespread.

Milk puddings and other dishes, for example, were common not only because they could be left to cook but to use up milk before it turned sour. All over the world, people ate within a fairly narrow range of food, dictated by the availability of ingredients. Baking, on the other hand, was a craft that required care and attention and for the exercise of which women cleared time, say, once a week.

Seasonality in a food-producing country also had its charms. People waited for the few months in the year in which some fruits and vegetables became available and relished them in their freshness; whereas some fruits, available around the year now, imported from overseas, are expensive, lack freshness and flavour, and quickly go bad.

Despite this much reviled Kiwi diet, it was nutritious enough for observers during the Second World War to remark that while New Zealand soldiers were not tall, they were strong and stocky and had marked resources of stamina. Perhaps the reason was that even if the cooking was not delicately done, food was abundant in New Zealand, even relatively so during the Depression, and milk, meat and potatoes were cheap.

The standard dinner until at least the 1960s was meat with potatoes and two or three usually overcooked greens. These meals reflected what was available. Lamb, beef and especially offal meats were cheap and plentiful and those vegetables easily grown in home gardens were fresh and available seasonally. Seasonality had a huge impact on what was eaten, not only vegetables but more especially fruit.

Potatoes, pumpkins and onion, turnips (or swedes) and to a lesser extent carrots and parsnips could be stored and were therefore available through much of the year, but marrows, green vegetables — cabbages (two varieties), lettuces (one variety), silver beet, dwarf (French) beans, scarlet runners and broad beans, green peas — had short lives out of the ground and could be ready for picking no more than three months in most parts of the country. Broccoli, asparagus, avocados, courgettes were rare, although courgettes, matured as marrows, were common.

Potatoes were eaten with every main meal and seasonality had an effect on them too. 'New' potatoes, small, white and freshly dug went with lamb and green peas (with mint) for many a Christmas dinner. During most of the year 'old' or 'main crop' potatoes were eaten, simply boiled and sometimes mashed, or baked in their jackets, or roasted with a joint of meat. Every home gardener grew potatoes, some enough to get through the year although they were so staple it required a lot of garden space to grow that many for a family.

Most main meals in households were constructed around meat and potatoes. Large joints of beef and lamb and big pots of boiled corned beef or mutton were cooked hot for a main meal. What was left over was sliced and eaten as cold meat with vegetables the following night, and the remainder minced and reused for shepherd's (or cottage) pie, or possibly savoury mince on the third night. In the days before refrigerators this was prudent use of meat. Fried (rarely grilled) lamb and pork chops, pork sausages and offal meats, accompanied by the usual vegetables, were the fill-ins.

Dripping was used for cooking and the intake of animal fats very high. For example, dripping would be placed around roasts which were themselves usually surrounded by their own fat. After the meat was cooked, the dripping left in the roasting dish was poured into a fat tin and kept for frying and the next roast. The effects of so much fat in the diet were reduced by walking and activity concomitant with daily work in the days before motorcars, farmbikes and wholly automated machinery. People of the time would have been enormously amused to be told that fifteen minutes' walking twice a week would reduce their susceptibility to dementia. Oh, that they could have been so lucky.

Offal meats were also basic and vigorously advocated by nutritionists for their basic food value. One of the most influential, physician, ZB network broadcaster and journalist during and immediately after the Second World War, Dr Guy Chapman, said: '...and remember that liver, kidney, tripe, heart and so on are really the most valuable meats that you can eat and, no matter how much money you have, they should be included at least once, and better still, twice, in your diet'.

Chapman even advocated using liver, kidney or heart twice a week to help prevent constipation, which was a common, indeed pervasive, complaint of the time and one on which he devoted many a radio broadcast. Sweetbreads, black and white puddings and brains were also regularly eaten, along with ox and lamb tongue and pig trotters. Offal is no longer easily available or cheap and has had a bad press from the medical profession in recent years because of its generally high cholesterol content, although its high levels of iron have been missed by many. The break from the past is dramatised by the fact that offal foods, once so commonly on display, were gradually removed from butcher-shop windows because the sight of them offended young shoppers.

Fish was a Friday thing, an imperative for Catholics but also a chance for a change of diet for the rest. Fish shops also sold poultry and rabbit. Fish was not as common in meals then as later, although fish-and-chip shops abounded. Rabbits were widely eaten, as much or more than poultry. Poultry was mainly bred and farmed for eggs, seldom used as a meat, and bore no resemblance to the specially bred, pumped-up meat birds common since the advent of supermarkets.

My own memories of food before the 1960s were that it was certainly plentiful. Breakfasts were hearty with cereal (usually porridge), toast and sometimes an egg. School lunches were rather skimpy on today's standards, but dinner was always plentiful and certainly rich in carbohydrate and protein. In smaller towns and cities in early years, families would gather for dinner in the middle of the day. Schools were often in the close neighbourhood and along with fathers' work a short walk or bicycle ride away from home. Later, as cities grew bigger, dinner at midday was impracticable, so it was shifted to as soon after 5 p.m. as father could get home from work, or after 6 p.m. if he went to the pub. Eating after 6 p.m. was considered a late meal. Three meals a day — with breakfast around 7.30 a.m., lunch around noon, and dinner around 6 p.m. — was the standard.

School lunches consisted almost invariably of lettuce, cheese and marmite/vegemite sandwiches with biscuits or piece of cake and a piece of fruit; and rare among my contemporaries I loved the free school milk. White bread was the norm; indeed brown bread was almost entirely white bread dyed.

## Chapter Eighteen

# Lamb Makes a Comeback

In the opening decades of New Zealand settlement by Europeans, wool grown on sheep grazing native tussock (and, later, English grasses) in the country's big paddocks was as valuable as the gold thread spun from straw by Rumpelstiltskin's slave. It was an international commodity requiring no refrigeration to move to markets around the world and, if its price came and went over the years, it stayed high enough for fortunes to be spun from it, mainly in Canterbury, Hawke's Bay, the East Coast and Otago.

That changed after the Second World War when synthetic fibres became more plentiful, cheaper and available for variable use. Synthetics had the market advantage of being able to meet the required specifications at prearranged prices, whereas wool was sometimes not consistently true to label when it arrived at the manufacturing plant, and prices fluctuated from week to week, affected, of course, by the levels of international production and consumption.

The wool-producing nations combined, with some success, to market wool under the imprimatur of the Woolmark which tried to guarantee even quality of product. Fine wool held up better than the coarse wools which were predominantly what New Zealand grew for the burgeoning carpet industry.

When Britain was subsumed by Europe, wool did not suffer the same sudden, disastrous consequence as meat and dairy products. But, as the competition from synthetics spread, the price settled at a level that was closer to the value of Rumpelstiltskin's straw than the gold spun from it.

By the 1960s, New Zealand sheep flocks included close to thirty breeds, among them animals that had come in originally from Britain, some that had been developed here, and others imported through Australia. The Merino had just held its own on the dry hills and high country of the South Island, and others such as the Cheviot, Corriedale, Halfbred, English Leicester, Border Leicester, and Southdown were still present in numbers.

But easily the most ubiquitous animal was the Romney, comprising near enough to three-quarters of the national flock, a successor early in the century to the hardy Lincoln. Imported as Romney Marsh from the 1850s, the local version developed into a type of its own and became the standard dual-purpose sheep with a good carcase and sound strong wool. Other Romney offshoots during the 1960s were making their mark — Perendale (Cheviot/Romney) on the hard country and Coopworth (Border Leicester/Romney) on the foothills and flats.

The Drysdale was also bred from the Romney, carrying a gene for hairy fleece and horns, primarily to provide exceptionally coarse wool as a component of carpets. It was a niche that had previously been filled by Scottish Blackface not available in this country.

As the second decade of the twenty-first century ended, sheep numbers were declining steadily. The best prospect for wool was the fine Merino knitted into quality garments for the high end of the garment industry.

The Southdown arrived in New Zealand early and was the most common terminal sire which brought early maturity to lambs and allowed them to lay down fat for the sort of blocky carcase beloved at the time by the British market.

The industry was set in its ways with Romney breeding ewes shorn for their wool and providing lambs for the meat trade. Trouble was the Romney had changed shape from the original imports and was having fertility problems and lambing percentages were usually a shade under a hundred. Then demand for fat lambs waned in favour of lean meat.

One industry executive says that in the early 1970s, when he was working in Britain, the carcase that won the Best New Zealand

In the early 1970s, 'Prime', with a covering of fat, was the grade that was tops in what was actually called the 'fat lamb' trade. Another grade, called 'seconds' did not have as much fat and, as tastes changed, the seconds began to fetch the higher price.

Lamb competition at Smithfield was a Southdown cross which had a layer of fat all over it. 'Prime', with a covering of fat, was the grade that was tops in what was actually called the 'fat lamb' trade. Another grade, called 'seconds' did not have as much fat and, as tastes changed, the seconds began to fetch the higher price.

About the time this was happening — as a consequence of falling meat prices in the wake of the EEC embracing Britain and few alternative markets opening up — successive New Zealand governments began increasing subsidies to sheep farmers, especially via Robert Muldoon's 'ewe retention scheme', to keep the industry going through what they perceived to be a cyclical downturn and an overpriced currency. The effect was an alarming increase in the national flock, from forty-seven million in 1960 to sixty million in 1970, and more than seventy million in the early 1980s.

Then, as the supports were withdrawn in the mid-1980s, sheep numbers plummeted — to fifty-eight million in 1990, forty-two million in 2000 and down to thirty-nine million in 2005. What was extraordinary and a tribute to farmers' acumen, foresight and efficiency was that those thirty-nine million sheep produced more export lamb, about 358,000 tonnes, than in 1987 when the total from more than sixty million lambs was just under 350,000 tonnes.

The reasons were a soaring lambing percentage — up from an historical figure of around 100 or slightly less in 1990 to more than 120 fifteen years later —and a big lift in the average carcase weight which moved up from twelve kilograms to seventeen in the twenty years to 2004. All because of the introduction of new, European breeds.

The sheep got so big so quickly shearers found them hard to handle, and investment in technology had not kept up for the industry to work towards a solution. This slowed the process and imposed heavy demands on the Accident Compensation Corporation and consequently on premiums. The problem was compounded by the ageing of shearers with a declining number taking up such a tough job.

When the National Government's supplementary minimum prices for sheep (SMPs) were taken off in the mid-1980s, farmers started looking at efficiencies and costs — and realised that the cost of taking the wool off a sheep was economically becoming less attractive than ever. Most of them figured lamb was going to provide a better base return for the costs involved, so the emphasis of their farming practice was directed towards getting more and bigger lambs from the declining number of ewes, with wool virtually a by-product.

The following fifteen years confirmed the decision was the right one for the time. It also gave farmers confidence that they could face world markets.

Interestingly enough, Ruakura scientist Dr Alan Kirton said in 1964 that New Zealand would have to increase its pastoral animal population by four per cent a year just to keep up, and he noted that the application of scientific principles of breeding in sheep and beef cattle were forty years behind the dairy industry. In fact, the economic productivity of the pastoral sector — not animal numbers — increased four per cent a year over the twenty years from 1985, and once it shucked off lazy practices, the sheep and beef industries were contributors to this performance. What gave lamb its productive thrust was the introduction of exotic breeds to boost flock fecundity and increase carcase weight.

The lift in lambing percentages followed the importation of, first the Finnish Landrace and Texel, both of which also added to the body size of progeny when used as terminal sires over traditional local breeds, including Romney. Romney still made up half the national flock in the new century. Texel and Finns both came from northern Europe. It was a revolution. Lambing percentages had for years been locked in around 100 per cent or a shade lower.

The Texel and Finn were followed by East Friesians (also a northern Europe breed) in 1996 and Dorper, a breed of South African origin, in 2000. The East Friesian was not only fecund but a prolific milker kept primarily as a milch animal in some countries. Other late introductions included the Gotland Pelt and White Headed Marsh. The Oxford Down was reintroduced in 1990 after disappearing in this country following original importation in 1906.

The first attempts to bring in the highly fertile breeds twenty years earlier had ended in disaster when scrapie disease broke out in a quarantined flock and all the sheep were slaughtered. A fatal disease that affects the central nervous system of sheep and goats, scrapie was feared because once in the national flock it would be almost impossible to eradicate, certainly if it became endemic to feral goats, for example.

Scrapie is carried by an agent smaller than the smallest known virus, and is even more feared nowadays than previously because it is a 'transmissible spongiform encephalopathy' (TSE) disease, similar to 'bovine spongiform encephalopathy' (BSE), popularly known as 'mad cow disease', which has a link with other species, including humans. No known transfer has occurred between scrapie-infected sheep and humans, but all TSEs are now considered threatening.

Even before mad cow disease had seriously damaged the rural economy of Britain, scrapie was considered a huge threat to the New Zealand economy. It is entrenched in the United States where effects on the export and domestic sheep industry have been so serious the government there decided to embark on an eradication policy, one that will take years.

However, by the 1980s, technology had triumphed with the breeds successfully imported into New Zealand either as genetic material or as quarantined animals, or both. The new infusions of genes resulted in a better than twenty-five per cent improved lambing percentage as well as heavier lambs.

The average lamb weight in the new century was climbing from about 12.5 kilograms to sixteen or even eighteen. Some farmers pushed for twenty-five but most accepted that the industry had developed a market for a distinct type of New Zealand lamb, ideal for cuts. Other countries, including Australia, and particularly the United States, produced larger animals. The common joke is American sheep are bred large so they can fight off the coyotes. So some New Zealand meat companies began discounting carcases where farmers had bred big lambs and held them just that little bit longer on grass to get added weight.

An influence on these decisions was improvements in chilled meat technology and in access to Europe for chilled as opposed to frozen meat. The voluntary restraint agreement with Europe (VRA) originally provided for more frozen than chilled meat, but that changed after New Zealand put pressure on to get access to the European common border. After that, individual countries that wanted to keep the lamb out had more difficulty arguing against it. Demand for the chilled cuts grew and that changed the profile of the consumer — from the poorer end which bought cheap frozen lamb to the more affluent which preferred the chilled cuts.

The American market was similar with no restraint on volumes since the aberrant bid during the Clinton presidency to impose quotas, a move stopped by the World Trade Organisation. The Americans, too, had begun by buying New Zealand frozen lamb and housewives were disappointed when they pushed it into the oven and it didn't emerge the same as fresh meat.

The New Zealand industry researched ageing and conditioning processes, but chilled meat gradually became the prime export. This market had always needed vigilance. For example, Americans enjoy lamb hot but seldom eat it cold, which gives an advantage to cuts rather than roast joints. It was also price sensitive. Once the price exceeded a certain limit, restaurant meals involving racks fell away, although some restaurateurs moved to lamb shanks and found them popular.

Demand for chilled cuts grew and that changed the profile of the consumer — from the poorer end which bought cheap frozen lamb to the more affluent which preferred the chilled cuts.

By 2005, most sheep and beef cattle farms were still operated by families on basic grazing systems which depended in particular on the nature of the land and climate they were situated on. Most were on hill country which ranged from moderate to steep. The grass was supplemented by hay, silage and in some cases fodder crops. Labour was provided by the farmers themselves, some casual labour and, increasingly, contractors.

Mobile phones kept everyone in touch. Farmers were becoming increasingly computer literate and spent productive time on long-distance purchases, price-checking and searching for good deals. Broadband reached into the provincial areas. It had all looked positive enough but, as the 2020s approached, sheep numbers were in slow but steady decline by two or three per cent a year to about 27 million. The densest sheep region was Manawatū-Whanganui. As with beef, a problem was a health-driven move away from meat in the traditional Western countries, although mutton was still a strong component of the Middle Eastern diet.

## Chapter Nineteen

# Bringing in the Beef

The prevailing angst among New Zealand farmers in the 1960s and 1970s was that foot and mouth disease would arrive and destroy an economy that remained so dependent on the export of pastoral products. But, at the same time, they wanted to find a risk-free way to bring in other cattle to spice up beef and dairy breeds.

Remoteness and long, slow journeys to get here had been historical allies against many exotic diseases. But direct contacts with the outside world grew in volume and speed. The first passenger jets brought in more people and freight more swiftly and regularly than ever before from the early 1960s. The tempo increased further a decade later with the arrival daily of jumbo jets carrying up to 400 people and large quantities of cargo.

Farmers and the Department of Agriculture — although wary of a number of diseases historically unknown in New Zealand — most feared foot and mouth, a highly contagious viral infection with low mortality in adult animals, although it may kill the young. Foot and mouth can spread through bovine species, sheep, goats, pigs and deer at a frightening pace. The incubation period is up to a fortnight, and if infection spread into feral animals, eradication from New Zealand would be a logistical nightmare.

The slaughter of infected stock and all stock that had come into contact with victims of the disease would follow an outbreak. All stock movements would cease, and no other country would take pastoral exports until New Zealand was declared free of the disease. Because of the possibility of reinfection, all slaughtered animals and any animal products would be burnt, along with any farm material or equipment that could not be scrubbed and disinfected. Foot and mouth remains endemic in many Asian, African, Middle Eastern and South American countries, and outbreaks are a constant threat to Britain and Europe, mainly through the smuggling of illegal meat.

New Zealand government departments trained many people — from veterinarians and farm advisory officers to police and Customs officers — during the three decades after the Second World War, first, to keep out foot and mouth and, second, to move swiftly to eradicate it before it could spread should it arrive.

But to overcome this dread of disease arriving on imported animals, techniques were developing as early as the 1960s to enable beef breeds to be brought in, even though the difficulty of detecting scrapie stalled bids for a while to bring in new sheep strains. The Dairy Board, through its Livestock Improvement Association at Newstead in the Waikato, was enthusiastically involved in devising ways of safely importing new cattle breeds.

The first beef animal in New Zealand was the Durham, an early type of Shorthorn, which arrived from Australia with missionaries in the second decade of the nineteenth century. Later-type Shorthorns were imported in 1849 and became popular because of their all-purpose role as a draught, dairy and beef animal, especially useful in remote areas where settlers were farming not far above subsistence level. The Shorthorn was superseded as a dairy cow by the single-purpose Ayrshire and Jersey but was present among beef herds for many decades.

During the nineteenth century, the Hereford and the Aberdeen-Angus (as it was called then) gradually took over the beef production role, ideally suited to both New Zealand's fertile lowlands and steep hill country. Of the two, Angus gained ascendancy, perhaps because it was better on the hills, able to survive comfortably and produce calves that put on weight while still on the slopes. But both are hardy and able to recover quickly from periods when grass is scarce. Herds in the modern beef-fattening industry are mostly made up of crossbreds from Angus and Hereford. It gets harder as the years go by to find herds of purebred Angus and Hereford breeding cows.

For all their anxiety about exotic diseases, farmers believed importing new exotic cattle breeds was important for the future development of the pastoral industry. New genes might expand the context of farming both by producing bigger and more fertile sheep and meatier prime beef animals.

One of the big moves in the 1960s was into dairy beef, using beef bulls as terminal sires over dairy cows which were surplus to requirements for dairy herd replacement. Large-framed European breeds might prove ideal for that job, especially with the growing spread through dairy breeds of the Friesian, a large-framed animal good enough on its own and ideal for crossing with beef sires.

Of the several exotic breeds brought in since the 1960s, most have been used, and very effectively, as terminal sires over dairy cows. The upgrading of dairy beef quality with the use of beef sires became a major campaign in the late 1960s and the use of new genes was part of it. British cattle had been developed for prime, fat carcases, whereas European animals had been bred for pulling wagons or ploughs and their heavily muscled, large frames provided lean meat.

The first to arrive in New Zealand, and it did so with a fanfare, was the Charolais, the big-rumped white cattle that originated in France but came here first from Britain in the form of semen in 1965. Sixty-one bulls followed between 1969 and 1981. Calving problems emerged early on, but they were gradually ameliorated.

Then the gene pool flooded with a tsunami of breeds from many parts of the world. After the Charolais came Maine-Anjou, Limousin, Salers and Blonde d'Aquitaine, all also from France; Simmental, a big-framed Swiss breed; Chianina, the tallest cattle in the world, Romagnola, Piedmontese and Marchigiana from Italy; Santa Gertrudis, a breed developed in Texas from Shorthorns and Brahmans; and Murray Grey, a breed fixed from Shorthorn and Angus in Australia.

In 1990, Shaver cattle were introduced from Canada where the 'composite' animal was developed from nine breeds. Surprisingly, most of these animals are now merely curiosities, although Charolais and Limousin are present in numbers and some of the others have their fans. The role of these exotics became mainly that of terminal sires for dairy cows and no doubts lingered at the beginning of the twenty-first century over their contribution to dairy beef production. By then dairy beef made up well over half of beef exports, mostly to the United States for manufacturing into hamburgers.

But none of the introduced breeds challenged the ubiquity on the hills and the fattening paddocks of the Angus and Hereford. These cattle had adapted well to the New Zealand environment and the farming practice of outside grazing that had prevailed over the years. Just over twenty per cent of national beef production was from the Angus, and about thirty per cent either Hereford or Angus–Hereford cross. The rest was dairy beef from Friesians or from crosses usually involving Friesians.

The stocky Angus arrived here first in 1863, six years before the first Hereford, and within a few years farmers were exporting the cattle to Australia. Although initially the Hereford was more favoured, the Angus caught up in the late 1920s and has since surpassed all other beef breeds. In the past few decades, Angus cattle and semen have been brought in from Britain and North America to refresh the locally adapted animal which has itself been exported to countries in North America, Asia, Europe and even back to Scotland where the breed originated.

Beef cattle were historically regarded as ancillary to sheep farming. They were used for pasture control on hills, trampling and eating rank grasses and shoots of fern, mānuka and other woody second growth, especially on land that was not carrying enough sheep to economically warrant the regular application of fertiliser. As one scientist wrote in the mid-1930s: 'A long and arduous experience in hill country management has … resulted in the utilisation of cattle as the best machine for pasture and second-growth control.'

Some farmers believed sheep prospered better sharing grazing with cattle not only because they ate the coarse growth but because they also ingested the larvae of internal parasites which infected sheep but failed to gain accommodation in cattle. Sheep farming and dairying were solidly profitable in the 1960s, but fattening specialised breeds for beef was not an enterprise that could exist economically on its own; so beef was essentially a by-product of sheep farming and dairying.

The specialised beef breeds were most densely farmed on high-country sheep properties in Hawke's Bay, Poverty Bay and the Wellington region. In other areas, the meat from culled Jersey and Ayrshire cows, which predominated in dairy herds, was low grade, suitable only for grinding and manufacturing for the hamburger business.

Beef cattle were historically regarded as ancillary to sheep farming. They were used for pasture control on hills, trampling and eating rank grasses and shoots of fern, mānuka and other woody second growth.

New Zealand has never seriously contested the world market for prime beef which comes from cattle finished with grain on feeding lots thus developing the 'marbling' of intermuscular fat that enriches flavour and has always commanded a premium in America, other traditional Western markets and in Japan. Australia, with cheap grain available, had grasped some of this prime beef trade. The top New Zealand beef, free-ranging and grass-fed, comes in below that and will continue to do so in the foreseeable future.

The farming of cattle for beef changed over the last decade of the twentieth century and became economically more important as markets opened wider in the opening years of the twenty-first. Breeding cows still remained effective at improving pasture quality. Mostly delivering a ninety-five per cent calving rate, they weaned calves at 230 kilograms live weight or better. In a growing number of cases, the young cattle were finished not on the hill country farms but on flat and fertile properties whose owners exploited the difference between buying price off the hills and the later selling price. Most animals were taken through for slaughter at twenty months and around a 300-kilogram carcase. How much money the finishing farmer made depended on the cost of feed and the prices on the North American market.

By the second decade of the twenty-first century, beef cattle numbers were down from 4.5 million to 3.6 million and falling. The totality of beef cattle, including surplus from the dairy herd, numbered more than 4.5 million. The biggest concentrations were in the Waikato, Manawatū-Whanganui, Hawke's Bay, Canterbury and Northland. Although New Zealanders ate about twenty per cent of the beef produced (as always, more per capita than lamb), export prices remained heavily dependent on the American market, which in turn was affected annually by the level of domestic beef production, and by the availability and price of pork and poultry, but producing beef was still a viable choice, given the expertise and the right sort of land.

The industry was helped, as lamb exports had been, by the improvement in chilling techniques, providing a much more attractive product than freezing. Chilled beef was exported to Britain in the 1930s, but the distance was great and handling distribution, obviously more urgent than frozen meat, was a constant problem. But with some prescience, an agricultural economics lecturer at Canterbury Agricultural College, Lincoln, Dr Ivan Weston, wrote in 1935: 'Any major expansion of the beef industry in this country is likely to depend on the success with which chilled beef can be marketed overseas. At present the marketing of chilled New Zealand beef is in the early experimental phase and its ultimate outcome cannot be foreseen.'

Too few beef cattle complementing sheep was then seen as a problem and Weston continued that more attractive returns from beef would result in: 'An improvement in sheep carrying qualities of the pastures ... when the roughage was properly eaten out. In such circumstances beef cattle would rank not only as mowing-machines but as direct income-earners as well and the farmer would, consequently, have a wider field over which to spread his risks.'

Quotas of beef and veal into the United States remained a problem and still carried a tariff. Beef export quotas to the European Union were increased threefold, although only to 1300 tonnes and at a twenty per cent tariff.

New Zealand beef farmers were exploiting every other opportunity. As dairy herds built up spectacularly after the late 1980s, some farmers began holding on to all their dairy calves and running them on leased hill country for sale as milking cows, and some beef farmers bought heifers in to run on their hills for resale later to farmers building up their dairy herds. Friesian bulls from dairy herds, fast-growing and with lean meat, also became available in large numbers and in some parts of the country replaced Angus/Hereford cattle on the rolling hill country and even on steep hills that the beef breeds had called their own for so long.

Beef, like most New Zealand pastoral products, faced the long-term difficulty of providing a product for developed countries with Western tastes, the very markets which financially supported pastoral industries of their own and erected barriers to keep cheaper suppliers out. New Zealand's share of world trade was not as great as with lamb and dairy products. Australia and a number of South American countries also produced large quantities, crowding the available international markets.

An abiding problem as the years of the twenty-first century rolled on was the pulling back from meat eating for health reasons in so many traditional markets, and a sense that beef cattle grazing on tolerably fertile farms with reasonable rainfall or artesian resources was not a highly productive use of the land.

## Chapter Twenty

# Feral Friends and Enemies

In twenty years from the 1960s, deer were transformed from feral pests to domesticated farm animals in an extraordinary demonstration of the economic adaptability of the rural New Zealander.

In many countries deer had been farmed, in the loosest sense of the word, for centuries — by nomads who followed migrating herds and lived off them, by Asians who kept them in enclosures, and by Europeans who confined them in forest parks for the aristocracy to 'hunt'. But in this country they were incorporated within a modern, intensive, grasslands farming system for the first time anywhere in the world.

The speed with which deer were taken from the bush, placed in paddocks behind high fences and domesticated to the degree that they could be controlled by farmers astonished many. And yet the cause was simple enough. An infrastructure of packing houses had been set up to service already developed export markets for venison created by hunters increasingly using helicopters to shoot deer in the bush and bring out the carcases for processing. They were so successful that feral deer numbers slumped and the surviving animals took to more and more remote country. With supply drying up, the simple solution was to capture deer, breed them on farms and provide a base for the industry.

The pioneers of the industry, led by Sir Peter Elworthy, were held to ridicule — deer were wild animals and would never be domesticated. But even the farmers themselves were astonished at how, within just one generation, deer took to farm conditions. Two things were important to the establishment of a secure industry. The first was that two markets already existed: one in central Europe for venison, and the other was in Asian countries, most particularly Korea, for deer antlers, which fetched high prices as an important ingredient of Asian medicines. Both these markets welcomed the prospect of increased supplies of product. The other factor was the determination of a few farmers to ignore the red tape that hampered the other primary industries and to find out for themselves what could and could not be done.

Red deer (*Cervus elaphus*) were liberated in New Zealand by British settlers from the 1850s. The idea was to provide game for hunters, and the big 'heads' that soon became common in this country lured wealthy British aristocrats to Otago and the Wairarapa, providing employment for several former Scottish gamekeepers. Deer increased in such numbers, however, that the system of licences being issued by then ubiquitous 'acclimatisation societies' was dropped and deer shooting became an egalitarian activity in New Zealand.

New breeds followed red deer: fallow, sika, samba, white-tailed, rusa and wapiti, as well as moose and gnus (African antelopes), neither of which survived. Those that did survive thrived in the congenial climate. Feed was abundant and the only predators were human beings with rifles, whose preference for stags with 'big heads' did little to lower deer numbers. And hunters were relatively few in a largely empty country.

By the first decade of the twentieth century, deer had become an ecological menace, causing large-scale erosion by stripping native grasses and shrubs from the flats and slopes of watershed valleys, and by eating leaves and shrubs, causing bush to degenerate. Their numbers were such that they seriously lowered sheep stocking rates on the high country and led to farmers calling for government action. Their complaints carried weight because wool was still important to the New Zealand economy.

The situation was regarded as so serious that a 'Deer Menace' conference was held in Christchurch in 1930 and for the first time all protection was removed from deer. The Internal Affairs Department mounted a campaign against what were declared 'noxious animals' and the task was given to G.F. Yerex, who became known for the rest of his life to all professional deer shooters as 'The Skipper'. Yerex recruited men used to the backblocks, especially high-country shepherds and hunting guides.

These hardy men became known in later years as 'cullers'— a misnomer given that their task was eradication not culling. They shot hundreds of thousands of deer, leaving their carcases to rot in the bush but were encouraged to take tails, pizzles and sinews to supply the Asian market. The department's primary aim was to kill as many deer as possible, but the hunters were well aware of the value of the skins, antlers, various by-products and even the venison if it could be got to the chiller in time. Such facilities were springing up in very remote areas and many government hunters took up deer stalking as a profitable enterprise. The profits were so good that jet boats and light aircraft become involved.

The export market for venison in Germany expanded steadily and good prices led to an increase in the infrastructure. The Vietnam War with its gunships inspired some entrepreneurs to use helicopters, so hunters took to the air. In the 1970s, they shot deer with automatic rifles, retrieved the carcases, delivered them to the nearest roadside point and then returned to the slaughter. The government was at last seeing deer numbers plummet at no cost to the taxpayer, and an export business was flourishing.

In their own way, the participants were freebooters, ignoring rules and regulations, and several helicopters paid the price — especially as competition among them heightened with deer numbers plummeting so fast it became hard enough to cover the cost of operating a helicopter let alone climbing insurance premiums. But when individuals realised the future lay in farming deer to improve availability of the produce at the right time for the market, the helicopters came into their own again, this time to bring them back alive with nets.

Businessmen who had financed the take-off of the export industry and some enterprising shooters and farmers then looked at the prospect of creating an orderly supply, putting captured deer behind the high-cost, two-metre-high wire-netting fences and breeding them. In the late 1960s, when some deer were already behind fences, pressure went on the government to formally authorise the farming.

Acting on the advice of a select committee of Parliament, the government came out in favour, despite opposition from some farming and environmental groups worried by the possibility that animals escaping from the farms would restock the feral herd and again threaten the native forest. The government's decision to approve the new industry was driven by an accepted need at that time for diversification of products and markets. The decision proved the right one with deer farming quickly becoming the largest new pastoral enterprise of the century.

Controlling regulations were constrictive at first but loosened as farming proved successful and associated problems minimal. Initially the deer were run on poor tussock land because that had seemed their habitat of choice in the wild. It didn't take farmers long, though, to discover that the deer had lived in rough country because they were safest there from hunters, that as with sheep and dairy cattle, they thrived best on good, open pastureland with, of course, silage, hay or grain supplements when required in winter.

No one anywhere had devised systems for running deer on grass in modern pastoral farming conditions so many farmers treated the farm as an extension of the bush. They killed their stags by shooting at between fifteen and eighteen months but the New Zealand reds had not reached target market weights by that time. Experience and common sense triumphed as practical management systems were devised, aided by research from the scientists at Invermay research station in Otago. Fence and yard designs, feed content and patterns, mating management, genetic selection recording, optimum economic weights for slaughter and other issues deeply understood by sheep and cattle farmers had to be worked out anew for deer.

Most farmers stocked with red deer but, to achieve market weights in time for the European Christmas trade, elk (wapiti), one of the largest deer species, were imported from Europe to produce yearling carcase and antler weights. In the late 1990s, stags with massive antlers were still being imported from Europe and Russia. The locally captured feral deer were gradually replaced by farm-bred stock which proved more tractable, although stags remained temperamental during the roar (mating season).

The government encouraged investment in the new enterprise as it was being established, which resulted in millions of dollars from city investors pouring in. Farmers throughout the country were cautious at first, but once the returns from deer were compared with those from sheep and cattle, there was a rush of interest. Other countries showed interest in what was happening and adopted New Zealand systems, but none had the high numbers of feral animals with which to build herds rapidly.

Several firms were exporting venison but were persuaded to co-operate in marketing. This led to the formation in the mid-1980s of the Game Industry Board (succeeded by Deer Industry New Zealand). In 1992, venison was rebranded as Cervena in a bid to clearly identify the New Zealand product, and promotion in the high-value European market was stepped up, especially in Germany, with its traditional affection for the meat, and Scandinavia. The much-vaunted analogy was that the especially tender Cervena from open New Zealand pastures was to venison what Champagne was to sparkling wine. The name, the promotional literature declared, came from a combination of *cervidae* ('deer') and *venison* (which originally meant 'hunting'). The new name was for a while heavily promoted within New Zealand in the hope of widening venison's acceptance domestically, especially as a healthy lean red meat.

> In 1992, venison was rebranded as Cervena in a bid to clearly identify the New Zealand product, and promotion in the high-value European market was stepped up.

While Germany had followed the New Zealand example and begun farming deer, its total of animals as the old century turned into the new was about one-twentieth of the national New Zealand herd, which numbered around two million — about the same as the combined total of all the other fourteen countries farming deer by then.

By the time the industry had settled down, herds numbered more than 4000, many run as secondary enterprises to sheep or cattle. Some specialised in venison while other farmers raised stags with antler size and formation that brought high prices. The farms were spread around the country but were most common in Southland, Canterbury and the Bay of Plenty. The main supplier of velvet (immature deer antler, harvested before it hardens and calcifies) internationally was Russia, but New Zealand became regarded as the second-best source.

The industry in New Zealand gained too much momentum in times of strong earnings, with hectares carrying deer jumping more than sixty per cent in the decade from 1994. The growth of production that went with that expansion had a major impact on world prices for venison, with exports jumping from 22,700 tonnes in 2004 to 27,431 tonnes in a single year. Most of this went to Germany, but slightly increasing volumes went to France and Sweden, and lower-value manufacturing meat went to Australia and Taiwan. Marketers were aiming at the United States as prices fell amid the growing volume of venison on sale in Europe. Deer farming profitability in New Zealand was in decline because of overproduction, and signs grew that some farmers were reducing their involvement.

The other product, deer velvet, was also coming under price pressure in its main markets in Asia. Depriving stags of their fighting antlers also helped manage them on the farm. The velvet renews itself every year and has a traditional market in Asia as an aphrodisiac and as an ingredient in a range of health products. The volume in world trade was affected by the fact that velvet harvesting was illegal in some countries, mostly in Europe, for animal welfare reasons. The tissue was taken in New Zealand using anaesthetics under veterinarian supervision.

By far the biggest buyer of New Zealand velvet was Korea, with varying levels of demand from China, Taiwan and Hong Kong. The market proved erratic, depending on fluctuating volumes of local supply, on the effect of Asian economic strength on discretionary spending, and on trade protection. In a bid to increase the demand for velvet, New Zealand producers, in association with AgResearch at Invermay, supported investigations into velvet as a dietary supplement promoting good health, enhanced athletic performance, sexual energy and acting as an agent for anti-ageing and immune system improvement. Claims, some extravagant, were being made before research had advanced very far.

Deer farming had settled down by the beginning of the twenty-first century and, although numbers were still declining on an annual basis almost twenty years later, the national herd still exceeded 1.5 million.

Farming another introduced pest, goats, never reached the level of economic importance of deer because no market developed that approached the value of venison demand in Europe. Although more goat meat was consumed in the world than sheep meat, almost all of it sold in Third World markets where returns have always been insufficient to attract New Zealand producers. Thus goat meat and goat dairy products remained a small segment of New Zealand's primary industry and were unable to shift a significant number of domestic consumers from strongly traditional wool, lamb, beef and bovine dairy foods and, therefore, were unable to develop an industry large enough to gain economic benefits from scale.

They enjoyed something of a boom during the sometimes frantic moves towards diversification in the 1970s and 1980s, but the industry was disorganised and fragmented and only in the 1990s did it gain a small but more rational place among the primary industries. In the twenty-first century some milking herds were providing milk for feta cheese which was earning a higher place in the New Zealand diet.

Goats have been domesticated farm animals for thousands of years and remain predominant producers of fibre, as well as of meat and milk in many parts of Africa and Asia. Both the meat and milk products are of proven nutritional value. Goat cheeses have substantial markets in Europe, especially in France and the Netherlands. But the advantage dairy cows and beef cattle have is their greater productivity.

Goats were introduced to New Zealand by Captain Cook in 1773 and 1777 and were later brought in in large numbers by whalers, sealers and the earliest settlers during the closing years of the eighteenth century and the beginning of the nineteenth. The intention was to domesticate some for their meat and to release others for hunting, not only on the mainland but as emergency food on many of the islands visited by whalers and sealers. Wild goats in their hundreds of thousands proved especially harmful to native bush and Department of Conservation hunters have shot vast numbers in a bid to keep numbers down.

With their hard mouths and voracity, goats proved useful also for keeping noxious weeds, especially blackberry, at bay on marginal land, and are still used for weed control.

Commercially, the most successful goat farming has been fibre: mohair, the fleece from Angoras; cashmere, the fine underdown grown by all goats (except Angoras); and cashgora, the fleece from Angora–feral crosses. Mohair is shorn twice a year from Angora herds mostly sited in the drier regions of the country. Most of the fleece is sold raw overseas where it is sorted and processed into products for the high end of the clothing market. Angora fleece is similar in composition to wool but more elastic with greater lustre and takes dye well. Its end-use is the fashion industry as well as luxury furniture fabrics.

The industry in New Zealand has had its small booms but too many busts, mainly because it was based on inferior breeding stock. Most of the world's supply is grown in South Africa and the United States (particularly in Texas), and new genetic material brought in from these two sources in the 1990s promised a more solid future for the industry based on improved quality.

China is the world's leading producer of cashmere, a yarn that also has its place at the luxury end of the world fibre market for clothing. Cashgora is a New Zealand product, a coarser fibre than mohair, which has found its own place in the textile industry.

Milking goats have been farmed in New Zealand for many years and gained a wider acceptance during the last quarter of the twentieth century, although the industry survived several periods of overproduction that forced some producers out of business and financially damaged others. The farmer-owned Dairy Goat Co-operative (NZ) Ltd gradually became the dominant producer and processor of dairy products, and like the bovine dairy industry was centred on the Waikato. In the 1980s, the co-operative developed what it claimed was the world's first infant formula based on goats' milk.

Most of New Zealand's goats' milk has been processed into powder and manufactured into consumer products, such as UHT milk, and exported, with heavy marketing emphasis on nutritional value, and it has supported a growing interest in feta cheese. Much more of the huge hunger for cheese in France is met by goats' milk than by cows' milk. In New Zealand this production has helped settle a well-organised industry with a stable base for future growth.

Goat carcase exports number about 135,000 a year, more than eighty per cent of them from feral or semi-feral animals mustered twice a year for slaughter. This represents around eight per cent of the world trade. Few farmers have concentrated on breeding up for meat with such a small market in New Zealand, but the importation in the 1990s of the Boer breed from South Africa brought some enthusiasm to the goat meat business. The Boer is a high-fertility, early-maturing, specialist meat breed, with faster growth rates than the crosses already in the country and produced good results from hybrid vigour when bred with local goats.

CHAPTER TWENTY-ONE

# Forests Fall and Rise Again

One of the major primary industries developed in New Zealand since the Second World War processes timber from a forestry resource that began in reafforestation schemes half a century before.

In the earliest days of European settlement, the indigenous timber resource seemed limitless and was the raw material for one of the first major extractive industries — along with kauri gum, the resinous sap found in the ground under the great kauri stands.

Kauri grew only north of a line from Port Waikato to Tauranga but the Northland forests covered the largest area. Until well into the twentieth century gangs of gumdiggers operating in Northland — many of them from the Dalmatian coast of Croatia (mostly then called 'Austrians' and later Dalmatians) — worked over the land probing and digging for the gum which was valued for its use in making varnishes and polishes.

Māori had long used the gum as fuel and for torches and in tattooing ink, and Captain Cook found some on the beach at Mercury Bay but thought it came from mangrove trees. First exports went to Sydney in 1815 and by the 1830s its value was established and Bay of Islands and Auckland merchants were buying it from Māori to send away. Farmers in Northland dug for the gum under the old kauri forests to supplement their income and the Dalmatians began arriving in 1885.

At the turn of the twentieth century, gum exports were earning an average of half a million pounds a year, about the same as flax and a lot more than cheese. Kauri timber was also exported in quantity from the earliest days because of its value in shipbuilding, with tall, straight, branchless trunks rising into the sky.

One of the most valuable of the native timbers was another native conifer, kahikatea, or white pine. Kahikatea, the tallest of the native trees, has a straight grain and long clean lengths. Because it was odourless and therefore did not taint food products, it was widely used for butter boxes, cheese crates and other food packaging. Until the Second World War, it was by far the most commonly exported timber, with rimu, valued for furniture, second. By the end of the war, kauri exports had stopped, and indigenous timbers ceased to be of export value.

Timber production from indigenous forests peaked in 1907 at 432 million superficial feet, spectacularly up from 261 million feet in 1900. Exports kept going up, though, until 1912, when ninety-five million feet was shipped out. Imports of timber — mainly Australian hardwoods for railway sleepers and telegraph poles — boomed during the first decade of the twentieth century to fifty million feet.

Deforestation by Māori and then European settlers, as noted in earlier chapters, severely reduced the indigenous forest cover. Alarm bells at the rate of deforestation were sounded from the 1870s. Authorities were becoming alarmed by the rate at which the forest cover was being slashed and burnt, especially in the North Island. In 1871, some provincial governments offered individual property owners four pounds for every acre planted with any kind of tree. In Canterbury and Otago, strong north-west winds made large shelter-belts attractive and many took advantage of the offer to plant radiata pine and macrocarpa (or Monterey cypress, *Cupressus macrocarpa*) plantations, as well as smaller areas in eucalyptus varieties. The Railway Department established small plantations of exotic trees during the 1880s.

Timber production from indigenous forests peaked in 1907 at 432 million superficial feet, spectacularly up from 261 million feet in 1900. Exports kept going up until 1912, when 95 million feet was shipped out.

In 1896, the big moves began when an afforestation branch of the Lands Department was formed and forest tree nurseries were set up at

Tapanui and Eweburn in the South Island and Rotorua in the North. The favoured tree became the exotic from North America, radiata pine (or Monterey pine, *Pinus radiata*), because of its extraordinarily rapid growth in New Zealand, maturing in not much more than twenty-five years, as fast as anywhere in the world. Macrocarpa remained a popular tree for single-line windbreaks.

The first major planting began in 1898 and ten years later plantings had reached 1000 acres a year. From the 1920s until 1936 came the planting surge (aided by relief labour from the unemployed during the Depression) that covered another 270,000 hectares — 150,000 by the state and 120,000 by private companies. In 1927 alone, 4.8 million trees were distributed from State Forest Service nurseries, mostly to individual private planters. Both state and private activity were involved. The largest plantings were in the central North Island. By the 1960s, claims were proudly made that Kāingaroa in the Bay of Plenty was the largest man-made forest in the world.

In 1961, twenty-five years after the end of the planting boom in 1936, state planting of exotics was increased again and the government dangled incentives in front of private companies to extend their forests as well.

By the end of the seventies, exotic forests covered about 750,000 hectares, mostly in the centre of the North Island, the site of the highest concentration of planting in the 1923–36 period. Generally the availability of undeveloped land dictated where the forests were located. The central plateau was a wise site for the province: the land in the region had for many years been considered marginal for pastoral farming because of an ailment called bush sickness that eventually led to serious unthriftiness in stock. After many years of research the cause was established as mineral deficiencies in the soil, particularly cobalt. This knowledge brought many thousands of hectares back into pasture, although the area remained host to the biggest forests and planting continued on marginal land.

After the Second World War, New Zealand Forest Products Ltd (then the country's biggest company) and Tasman Pulp and Paper Co. (formed in 1952) built an industrial superstructure over the burgeoning production of exotic timber from the maturing forests planted in the twenties.

The production of pulp and paper from the central North Island forest had been talked about since the earliest years of reafforestation but the capital required was huge, and the industry cyclical and internationally competitive. One investigation into paper production in the early years of the twentieth century took very little time to dispense with the idea because the amount of capital needed was then prohibitive.

In the early 1950s, the town of Kawerau was set up in the Bay of Plenty to serve the establishment of a nearby plant for pulp and newsprint production. Thus forest products of many types supported a major industry in the immediate post-war period. The foresight of a couple of generations of governments and foresters had provided the country with an asset of enormous value.

Farm forestry schemes subsidised by the government spread the area of forests, and the replanting of trees in hill and high country helped cut the damage by erosion which seriously affected regions like Poverty Bay in the post-war years.

Less than one-quarter of more than six million hectares of indigenous forest — mainly beech, kauri, rimu, tawa and taraire — remained in private hands, with the rest owned by the government and managed by the Department of Conservation. Legal provisions controlled how private owners harvested and sustainably managed their forests.

During the last half of the 1970s private plantings of exotics were outstripping those by the state for the first time. Many farmers combined livestock farming with forestry, taking advantage of government incentives to do so. By the 1980s, sheep and beef cattle farms had over 100,000 hectares in exotic trees. The timber mills were sited within the region of the forests, which were progressively milled and replanted in a sustainable circuit. The harvesting and processing of timber required such large amounts of capital that two major industrial corporations gained control of more than a quarter of the forests.

Plantation forests covered about 1.8 million hectares as plantings continued in the twenty-first century, with more than 100,000 hectares established in 2003, followed by a fairly steep decline. The government remained involved. Plantings included the East Coast Forestry Project set up by the government in 1992, originally to plant 200,000 hectares of commercial forest. The project was severely modified but about 40,000 hectares was planted on land at risk from erosion.

Government grants were available to assist landowners with the establishment and management of the forests. The Crown Forestry Group, on behalf of the government, managed twenty-five exotic forests spread over nearly 40,000 hectares, sixteen of them leased from Māori landowners.

Total exports were still more than $3.2 billion early in the twenty-first century. Some progress was being made towards the ideal of adding value by higher levels of processing, with exports of sawn timber and finished products, mainly to Australia and the United States, increasing in value. But the sale of logs to Korea, Japan and China remained the biggest single end use.

A new government elected in 2017 announced a plan to plant a billion trees in a decade, which would create an economic asset and provide jobs in declining rural regions and also counter climate change.

CHAPTER TWENTY-TWO

# The Earth Yields Abundant New Fruits

No doubt existed from the time the first tree sprung its crop that northern temperate-zone fruits would thrive in New Zealand. Missionaries brought the seed in and Māori, with no experience of trees that bore so abundantly, rapidly took peaches around the North Island in the manner of the mythical Johnny Appleseed in the United States. Wherever they went they sowed the seeds.

The director of the Department of Agriculture's horticulture division, J.A. Campbell, wrote in the mid-1930s that 'before organised settlement took place and for many years subsequently, groves of stone fruit trees, particularly peaches produced from stones planted by the Maoris, were found bordering practically every watercourse in the North Island. These trees bore heavy crops of beautiful fruit, and continued to do so until age and the ravages of introduced diseases eventually wiped them out.'

The traditional temperate-zone stone and pipfruits were fundamental to the domestic industry here for a century, and over most of that time, apples and pears dominated a steadily built-up export trade — until the kiwifruit and grapes bonanza of the 1970s and 1980s which brought about a revolution in New Zealand agriculture unmatched since refrigerated shipping.

Kiwifruit and grapevines, carried on wooden frames in long, straight green lines, redecorated countryside once dotted with sheep and cows. Both did well in a variety of regions, but kiwifruit grew more plentifully in the Bay of Plenty than elsewhere and grapes thrived especially in Marlborough and the sun-dried east coast of the North Island.

Kiwifruit originated in China but were reincarnated into an international delicacy in New Zealand. Grapes are as old as civilisation in the Middle East and Europe but some varieties found a new, reinvigorating home here for the making of wine.

In the early days of settlement, the stone and pip fruits were the staples and, with biosecurity unknown in its modern sense, immigrants continued to bring from other countries plant material on which diseases could easily survive the long sea journey. Codlin moth, woolly aphis, scale pests, apple and pear scab and a range of fungal diseases were soon rife and it was not until the Orchard and Garden Pests Act 1903 that action was taken to curb the spread. Old and neglected orchards were destroyed and a ban placed on the sale or distribution of infected fruit or trees. Two years later, arsenate of lead became a standard and effective control for codlin moth.

The earliest orchards in the sense of organised planting of fruit with the intention of harvesting for sale were sited on Auckland's North Shore in the 1850s. Orchards tended to be planted close to concentrated settlements of people in towns and cities because of the perishable nature of most ripening fruit before chilling facilities were available.

Unlike today, peaches, nectarines, apricots, apples, pears, plums and cherries had their brief seasons of high availability in summer and autumn during which consumers gorged themselves while they could and then waited in anticipation of fructuous times the following year.

The same applied to tropical imported fruit. Generations of New Zealand children received an orange in their Christmas stocking, a rare and exotic gift. Before the Second World War, lemons were grown on nearly 1000 acres in New Zealand and oranges on 400. By the end of the century, mandarins, oranges, lemons, tangelos and grapefruit were grown on more than 2000 hectares. Most of these fruits are subtropicals and do well here only in the mildest regions of New Zealand's temperate climate: Northland, Auckland, Bay of Plenty and the East Coast. Imports from Australia and the United States continued to bridge seasonal gaps.

The first apples and pears, like so many plants and animals, arrived with Samuel Marsden, who planted trees at the Church Missionary Society's Kerikeri base in the Bay of Islands. Twenty-five years later, the society reported that the pipfruit trees were producing abundantly at most of their stations. The development of an export trade was inhibited for many years by the perishability of fruit, except that apples and pears survived better than the soft stone-fruits.

Orchardists sent the first export consignments of apples to Chile in 1888 and Britain four years later. From the first decade of the twentieth century, a trade built up with South America which took 68,000 cases of apples in 1914 before the First World War brought it to a temporary halt. By this time, the main apple and pear production was moving inexorably towards Hawke's Bay and Nelson, still major sources of supply for both the domestic and export markets. Apple growers had discovered they could make a living from land that was marginal for pastoral farming — often land that had suffered severely from the depredations of rabbits — as long as the weather was right. Peaches found their most salubrious home in Hawke's Bay, and apricots in Central Otago.

Over the years, apple and pear orchardists followed generally in the footsteps of other primary producers towards co-operative distribution and marketing, helped by the reigning governments. A Fruit Control Board was put in place in 1924. It controlled all exports under one label, and four years later shipped a million cases.

Fruit and vegetable growers have always faced a more precarious existence than those involved in traditional New Zealand agricultural pursuits. Contrary weather — frosts, hail, heavy winds, floods and bird strike — could eliminate crops, or spoil them for export where high quality standards had to be met. While some technological developments helped ameliorate frost and bird damage, fruit-growing remained a risky business. Equally, overproduction in good seasons could corrupt markets, especially when it matched gluts from overseas producers.

On occasions over the years, the New Zealand industry reacted to economic forces to reform itself. In the decade after the First World War, orchards got bigger and fewer, with about 7500 growers on an estimated 50,000 acres shrinking to fewer than 6000 on about 27,000 acres. And again in the 1950s, the industry was persuaded to cut back the number of apple varieties from about 140 to fifteen and pear varieties from more than forty to nine.

> During the early years of the Second World War, apples were sold cheaply, given free to schoolchildren and consumption reached 265 apples a year for every New Zealander.

In more recent years, the number of varieties of apples grew again, although some varieties faded in competition with such desirable new commercial brands as Gala, Royal Gala, Fuji and Braeburn which soon made up more than seventy-five per cent of exports and on which international promotion was focused. Jazz apples entered the market in the new century. Except for the Japanese variety, Fuji, the new apples were the results of research and development in New Zealand where scientists were always striving for a better apple and pear.

During the early years of the Second World War, apples and pears were crowded out of the limited shipping space between New Zealand and Britain and the crop was sold domestically. The government subsidised growers when the price slumped below the cost of production. Apples were sold cheaply, even given free to schoolchildren and, according to one account, consumption reached 265 apples a year for every New Zealander. Substandard fruit was used as pig fodder.

The government in 1948 set up the Apple and Pear Board in much the same shape as the other producer boards, with grower and government representation. Subsidies to growers were removed and the board operated a guaranteed price to level out growers' returns between fluctuating prices over the seasons, a mechanism that lasted until the 1980s. In 1952, the board was given the power to control all purchasing and marketing.

The year the board began exporting to the United States, 1959, international production was increasing and squeezing prices, and also forcing up the standard of quality required. The move into the United States was a bid to widen the product destination which had for decades been predominantly Britain. At this time too, the growers began using new techniques, successfully experimenting with pruning methods and smaller trees — some with central leaders in pyramid shape, some as espaliers on wire frames. Combined with more effective disease and pest control, these developments gradually doubled the yield per hectare. More efficient storage and distribution systems kept apples on the domestic market virtually all year.

As the 1960s moved into the 1970s, the board launched a major marketing drive to diversify away from Britain and the five or six other established markets. Eventually New Zealand apples went on sale in more than forty international markets. From the 1960s, as the demand for higher quality rose, surplus apples that didn't make the export cut were used to develop special consumer products such as apple juice (combined also with other fruit juices) and bakery trade items such as apple pies and apple slices.

This ancillary processing industry also helped absorb excess production in years when international markets were jammed with apples. Products included a range of apple, pear, carrot and berry-fruit juice concentrates. Through an Australian-based subsidiary, Fruitmark, the New Zealand industry also sold concentrates, dried fruits, fruit pastes and dehydrated fruit and vegetable products.

In the 1980s, the board ended its guaranteed price scheme and adopted a bold strategy similar to that of the Dairy Board, buying as well as selling fruit from around the world, trying to command the trade as a kind of international broker.

Meanwhile exports reached more than seven million cases in 1984 and doubled to fourteen million cases only seven years later. This expansion of sales occurred without the ability to export to Australia since 1921, when a ban was invoked on the grounds that imports would introduce fireblight, a tree disease endemic in New Zealand. New Zealand claimed widespread scientific support for its case that fireblight could not be carried on picked fruit. The World Trade Organisation agreed and Japan, which had also imposed a ban on New Zealand apples, decided to comply with the WTO ruling.

However, for Australian apple growers immunity from competition was powerful inspiration for keeping pressure on their government not to relent on the ban even after eighty-five years. The issue remained unresolved until the second decade of the twenty-first century when Australia began to accept a small volume of exports under strict biosecurity conditions.

In 1991, the Apple and Pear Marketing Board had decided to brand its apple and pear exports under Enza, a name chosen, said the board, because it was short and punchy and clearly evoked New Zealand. International competition was heating up at the time, and became increasingly intense as production grew, especially in the southern hemisphere, which posed a seasonal challenge to New Zealand.

In difficult times, dissension grew among growers, despite the fact that the board was owned and governed by them as a co-operative. Enza had expanded too far as a bureaucracy, according to many, who insisted it was costing them too much money, even though it funded a lot of research, including work on new varieties. Many companies were trying to export apples and they applied pressure against the single buyer-seller concept and helped towards the imminent demise of the board.

The board's control ended and the industry corporatised with the passing of the Apple and Pear Restructuring Act at the end of 1999. The Act turned the board into a limited liability company, to be known as Enza Ltd. Shares were distributed among growers in proportion to supply levels over a period of years. Enza became just one of a number of selling and distribution players. The industry entered a period of disarray as it tried to settle into the new structure — or lack of structure to be more precise. Export companies competed against each other in world markets.

At a time of an international glut of apples and with kiwifruit and grapes enjoying steadier growth, the area of land in apples declined. Trees were still being pulled out in 2005 as apples continued to concede land to grapes in Nelson and on the east coast of the North Island, and the area in apples had already slumped from 14,541 hectares in 1999 to 11,715 in 2003.

The troubles came to a head in 2005 when sales in a Europe overloaded with apples slumped and so did the price, leaving growers and packers in New Zealand with deep financial problems. Faced with a calamity, twenty-nine growers and exporters agreed to form a market panel to administer a New Zealand pipfruit brand, complete with logo, using a quality imprint, or trustmark.

The task that lay ahead was to convince all those involved, especially exporters (and there were more than 100 of them), that demonstrable benefits would be gained from joining the panel and accepting its constraints along with its marketing plans.

The efficiency of the New Zealand industry had long been admired, and although the country was distant from markets it still had advantages: yield of apples per hectare was matched in the world only by the Netherlands; and scientists were leaders in the business of developing new consumer-attractive varieties, such as Jazz. So by 2017, the country was exporting record volumes, earning more than $750 million a season, still mostly to the UK, US and Europe but by then the volume to Asian countries was approaching half.

The kiwifruit story buzzed with glamour when it broke at the end of the 1970s, enhanced by the fact that the growers who had risked all on the future of the brown furry fruit became millionaires in a fairly short time on relatively small areas, most famously on No. 3 Road, Te Puke. Within a few years, growers in other places followed suit and the fruit supplanted apples and pears as the biggest horticultural export earner; so a man called Jim MacLoughlin and those who followed his example did much for the New Zealand economy as well as for themselves.

The kiwifruit plant, a Chinese native, had been transported around the world by curious travellers and botanists since the nineteenth century, but it took New Zealand orchardists to recognise its broad appeal and commercialise it. The fruit was then called Chinese gooseberry because of a vague similarity in appearance to the gooseberry when cut open and not from any similarity in outer appearance or flavour. The plant held a place in the Royal Botanic Gardens, Kew, from the mid-nineteenth century, and early in the twentieth, a British nursery firm, James Veitch and Sons, listed *Actinidia chinensis* under 'Novelties for sale' but apparently sold only male plants. All the buyers got was a vine with nice leaves and flowers.

In 1904, as the plant was being disseminated around the world —often by missionaries and more as a curiosity than a commercial prospect — Miss Isabel Fraser, principal of Wanganui Girls' College, arrived home from visiting her evangelist sister in the Yangtze Valley, in China. She brought with her some Chinese gooseberry seeds. They were grown by a Wanganui horticulturist and by 1920 a number of nurserymen were growing and selling the plants, among them a man called Hayward Wright. He developed the Hayward cultivar, which eventually became the variety on which the export industry came to rely.

Miss Fraser, by the way, was no ordinary traveller. She was a strong-minded feminist who became the founding principal of Iona College, a Presbyterian boarding school for girls in Havelock North. She worked for the first five years without pay to develop the school. The sister she visited in China later returned to New Zealand and joined the Iona teaching staff.

Chinese gooseberries became popular on the domestic market as orchards in Wanganui, Northland, Auckland and the Bay of Plenty began producing reasonable volumes of the fruit. By 1948 demand was small compared with tree tomatoes. Indeed as late as 1965, 184 acres were under kiwifruit and 377 under tree tomatoes. However, Te Puke orchardist Jim MacLoughlin invested his faith and his money in his belief that kiwifruit, sweet-fleshed and good keepers, had the potential to become internationally desired.

MacLoughlin made his first commercial planting in 1934, had seven acres planted in vines by 1940 and forty-five acres by 1955. He pushed to export the fruit, teaming up with another Te Puke grower. They shipped out 2000 cases in 1952 and the trade grew from there. MacLoughlin was later tagged 'Father of the Kiwifruit Industry'.

'Chinese gooseberries' was never considered a name that would resonate among consumers and was especially unappealing in the United States. One suggested name was 'melonette'. Then a copywriter at an advertising agency, Christine Cole Catley — later a well-known

'Chinese gooseberries' was never considered a name that would resonate among consumers and was especially unappealing in the United States. One suggested name was 'melonette'.

journalist, writer and publisher — submitted a list of thirty possible new names to her agency's client, exporters Turners and Growers. On the list were 'loveberries', 'fuzziberries' and 'kiwiberries'. Jack Turner liked 'kiwiberries' best when the names were put to him in 1959. He consulted a botanist who said they were not berries; they were fruit. Thus kiwifruit became the generic name internationally, including in China. Turner later lamented the name was not protected but doubt prevailed that a combination of two common words could have been registered anyway.

By the 1970s, New Zealand growers could not meet the demand and horticulturists in other countries welcomed the chance to buy the highly successful Hayward plants to compete. Orders flowed in to New Zealand nurseries from Australia, Italy, South America, South Africa, Israel and other countries. One prominent plant nursery firm accepted an order to sell to a French agent 10,000 Hayward plants a year for ten years.

An argument raged over whether the country was wise to sell the plants. Some claimed other climates would make it hard to achieve top quality; that it would be years before foreigners would be able to learn the idiosyncrasies of the plant; and that when they did New Zealand would still have a seasonal advantage. Others, less sanguine — and they turned out to be right — claimed foreign production would expand rapidly, compete in European markets with the New Zealand product and affect prices. Forty years after the plants were sold, Italy grew more kiwifruit than New Zealand, although New Zealand still controlled much of the international trade.

The number of hectares in kiwifruit in New Zealand grew from fewer than 10,000 in 1982 to just on 18,000 three years later and peaked at nearly 19,000 in 1988 before beginning to decline. Production, however, steadily increased on smaller areas and by 2004 the area sown was below 11,000 hectares again. In the 1970s and 1980s, pioneer growers like Jim MacLoughlin, centred on Te Puke's No. 3 Road, were generously rewarded with profits for their foresight, their dogged persistence and their flair.

New Zealand experienced a kiwifruit frenzy in the boom years of the 1980s as farmers converted land to grow the fruit and city investors found a way to make profits and avoid tax. Gradually the industry groped its way to sensible joint methods of processing, distribution and marketing and, consequently, reasonable profits were achieved in most years. But when the crop jumped from twenty-nine million trays in 1986 to forty-six million the following year and multiple agents were selling, the market staggered. Some growers thought the end of the golden days was in sight and got out of the industry, but most began to think of a single marketing entity in the model of the New Zealand Dairy Board.

That happened in 2000 when Zespri Group Ltd was formed, a single export entity with an integrated supply and delivery system, wholly owned by growers. Kiwifruit would henceforth be marketed under the Zespri brand, Zespri Green and Zespri Gold, causing some commentators to joke that a kiwifruit is a kiwifruit only when it is produced by Kiwis. Zespri Group Ltd immediately became the world's biggest kiwifruit marketer. By 2004, export sales were approaching $700 million, representing almost thirty-five per cent of New Zealand horticultural exports, well ahead of apples and pears.

In 2000, too, Zespri Gold — a soft, yellow-fleshed variety — was launched commercially. It had been developed over eleven years at the Te Puke Research Orchard and this time plant variety rights were taken out internationally. Trial export consignments had been shipped in 1997 and seven years later it was achieving market successes enough to encourage many Zespri Green growers to make a shift to gold. The fruit was already widely grown in New Zealand and, under licence, in a number of foreign countries, including Italy, France and Chile. By 2017, Zespri exported to more than fifty countries and earned $1.6 billion.

Kiwifruit and apple exports were worth more than fifteen times as much as all other fruits together, and faced no immediate challengers to their pre-eminence. Avocados, originally from South America and worth around $30 million a year in export returns in 2005 (mainly from Australia), were first grown commercially in New Zealand in the 1930s. Production increased by eighty-five per cent in the year to 2005 mainly because of a rapid expansion of plantings over the previous five years. A tropical fruit, avocados need care and attention in this temperate climate, sensitive to frost and wind and preferring free-draining soils. They occupied 550 hectares in commercial orchards.

The highest per capita consumption of avocados in the world is Mexico (more than eight kilograms) where it is the main ingredient of a national dish, guacamole, but consumption increased rapidly in the 1990s in the United States with its growing population of Latinos.

Two other fruits of South American origin, tamarillos and feijoas, were relished locally but are insignificant export earners, depending almost entirely on the US market.

Tamarillo trees are soft and need complete shelter from wind, while feijoas are delicate fruit and susceptible to frost. High hopes were held for tamarillos about the time kiwifruit were making an impact, but export returns have never reached great heights. They carried the name 'tree tomato' from the time they arrived in New Zealand in the nineteenth century until 'tamarillo' was coined as a more marketable alternative by Mr W. Thompson, a member of the New Zealand Tree Tomato Promotions Council, in 1967. Thompson joined the Māori, 'tama', which he translated as implying 'leadership', with what he thought was a Spanish ending. New Zealanders seem to have a gift for nomenclature because the name took and is now used in most parts of the world.

Tamarillos are widely grown in gardens in many other countries, but they have never achieved mass appeal. Commercial crops are not numerous, perhaps because the fruit has a tart rather than a sweet flavour.

Persimmons have been grown commercially in New Zealand only since the 1970s, but production built up strongly during the 1990s as the fruit proved profitable and in export demand, mainly from Asia, where the plant came from originally.

Grapes grown for wine seldom contest for land with the subtropical fruits. Like apples, grapes were first introduced by Samuel Marsden when he arrived here in 1814 with a variety of animals and temperate-zone plants to provide food for missionaries and Māori. James Busby, the New Zealand Resident, who was the first local vestige of British government, arrived in 1833 and planted vines at Waitangi. He had studied grape-growing and winemaking in France and written a book on the subject. When he was in New South Wales before coming to New Zealand, he bought land in the Hunter Valley, now a famous winemaking area, with the intention of growing grapes and making his own wine.

When he was appointed Resident, Busby was in Britain and before returning to the colonies he visited France and Spain collecting vine cuttings which he distributed through New South Wales, South Australia and New Zealand. By 1838, Busby had extensive vegetable gardens and a vineyard and was making wine which he sold to British troops stationed in the Bay of Islands. Within a few years, French settlers at Akaroa had their own vineyard, and French Catholic missionaries began winemaking in Hawke's Bay in 1865 at a place called Meanee, still a winemaking centre 140 years later.

A viticultural centre was set up by the government at the end of the nineteenth century at Te Kauwhata to import and investigate grape varieties and experimentally make wines. But despite bouts of enthusiasm for New Zealand as a potentially world-class site for growing grapes and making quality wine, no momentum gathered for the birth of an industry.

One reason may have been the British tradition of drinking beer, which many of the puritan population thought created enough social problems as it was. Few New Zealanders had acquired a taste for wine. The prohibition lobby was a powerful force within government and would certainly have discouraged prospective investors in a wine industry. One pocket of production, though, was in the north and west of Auckland among Dalmatian and Lebanese immigrants who brought with them a tradition of winemaking, one that survived the twentieth century.

The government viticulturist and oenologist for many years, Francis Berryman, wrote in the 1960s, about the time cheese-makers were daring to try camembert at Ōpouriao: 'Although the New Zealand climate is recognised as being particularly well suited to the production of table wines, the people, like most who are of British descent, prefer the stronger sweet wines, such as sweet sherry, port,

Muscat and Madeira types.' He was right as any New Zealander of that time would remember. Sherries and ports were still being consumed in large quantities and the national attitude towards table wines as the drink of effete Europeans was only just starting to fade.

Then, astonishingly, within twenty years, table wines and Continental cuisine became all the rage and winemakers of skill and taste were producing white wines acceptable to sophisticated drinkers anywhere in the world. Berryman had written in the 1960s: 'Up to the present the high cost of grape production … and the traditional bias of overseas wine markets have discouraged concerted efforts to develop a wine export trade.' That also changed dramatically.

In Berryman's time eighty-five per cent of the vineyards were in Auckland (425 acres) and Hawke's Bay (387 acres) and that was the way it stayed for some time. An earlier oenologist, Romeo Bragato, had found that the European *Vitis vinifera* varieties of grape predominated in Hawke's Bay but in Auckland Franco-American hybrids, including *Vitis labrusca*, were most common because they did better in higher rainfall regions and were largely resistant to phylloxera which was present, and deadly, in *vinifera*. It was some years before the lower yield but higher quality (for wine) European vines again became important, mostly grafted on to phylloxera-resistant rootstock.

The Department of Agriculture knew from the end of the nineteenth century that the best places to grow wine were Central Otago and the north of the South Island as well as Hawke's Bay. In 1895, Bragato, born in Yugoslavia and trained in Italy and France, came to New Zealand as a consultant to the government and reported that Central Otago was an ideal site for growing grapes for wine. 'There's no better country on the face of the earth for the production of Burgundy grapes,' he wrote. He also suggested Canterbury, Nelson, the Wairarapa, Hawke's Bay and some places in Northland had excellent potential. He did not visit Marlborough that time but was convinced of the value of the region at the top of the South Island.

Bragato returned to New Zealand in 1902 to set up the experimental station at Te Kauwhata for the Department of Agriculture but left in 1909, disillusioned by lack of government support and the general indifference of New Zealanders to wine. Not only was the department focus firmly on the pastoral industries, government advisors had decided to leave winemaking within the regions in which it already existed and not to spread it thinly, as one of them described it, around the country. It took private interests to sniff out the best vineyard sites much later in the century.

The powerhouse of the industry by 2005 was Marlborough. The first vines were planted there in 1973 but thirty years later, fruit trees were being pulled out and sheep shooed off the grass so land could be put down in grapes. Just under 10,000 hectares were in grapes by 2005. The area had grown by more than 1000 hectares a year since the new millennium and was five times larger than in 1996. Alluvial loams over gravelly subsoils in the Awatere and Wairau valleys, the longest annual sunshine hours in the country, cool nights, and a long growing season all contributed to the success of the region. The big grape there was Sauvignon Blanc, followed by Chardonnay, Pinot Noir and Riesling.

Neighbouring Nelson's area in grapes had grown from ninety-seven hectares to nearly 600. Over the same period, Central Otago had been developed and was producing delightful vintages, including Pinot Noir, and some new niches had been found to confirm the assessment of Bragato 100 years before. Southern Wairarapa began to produce lovely Pinot Noir. The area planted there doubled in three years to 758 hectares. The whole country was in the grip of the grape.

Some primary production traditionalists, knowing the history of booms and busts, became concerned at the rate of investment of land and money in grapes but wine export volumes and returns in the largest markets held up well over ten years. Britain and Ireland bought most, followed by the United States, Australia and Germany.

In 1960, no one could have foreseen the transformation of the fruit-growing industry in New Zealand. Although export earnings from pipfruit, kiwifruit and wine were still dwarfed by dairy and lamb returns, the explosion of horticulture gave some breadth and balance to the country's export profile. By 2017, 38,000 hectares were growing grapes for wine, about two-thirds of them in Marlborough.

CHAPTER TWENTY-THREE

# The Vegetable Business

Canning; advances in chilling and freezing and in packaging materials; the advent of supermarkets; air freight and faster transport on better roads; and the demise of the home vegetable garden — these changes transformed the vegetable and fruit industries in New Zealand after the Second World War.

The keeping qualities of some varieties — pipfruits, kiwifruit, onions, squash and potatoes among them — ensured an export market even in the days of relatively primitive packaging and shipping, but stone-fruits and many vegetables had to wait until processing could give pause to their rapid, natural deterioration.

James Wattie was a Hastings boy who studied accountancy as he worked in companies associated with the fruit, vegetable, arable and meat industries. Canning was in no way new to New Zealand, but Wattie saw prospects for preserving Hawke's Bay's prolific ability to grow stone-fruits, particularly peaches, and vegetables such as tomatoes, peas, beans and corn.

He was a pioneer. He formed J. Wattie Canneries in 1934, set up a factory in Hastings and gradually built up the country's largest canning business. As the Second World War ended the company was building up to produce a full range of canned and bottled foods: tomato and other soups, canned fruits and vegetables, jams, and the famous tomato sauce which became one of the most ubiquitous items in grocery shops and dairies.

Although Wattie's bought a farm in Hawke's Bay in 1945, it was soon processing enough fruit and vegetables to encourage market gardeners and some farmers on the fertile Heretaunga Plains around

Hastings and on the Gisborne flats to move into cropping on a larger scale than ever before. Apart from a wide variety of vegetables, apricots from Central Otago and berryfruits from other regions soon found a market for canning and freezing. In 1951, Wattie's — as the company became universally known — established a factory in Gisborne after persuading farmers there to plant sweetcorn, and more than four million cans of corn came off the assembly line in the first season. In 1970, a factory was set up in Christchurch, mainly to quick-freeze peas.

Wattie's was not the only company that took up canning and freezing food during the post-war period, but it was the biggest and best known until it was bought by the international food processing giant H.J. Heinz in 1992. Lever Brothers also began canning in Hastings and in Motueka in the 1950s, became a big producer of frozen and dehydrated foods, and was also an early producer of frozen meals, which followed quickly thereafter, and became especially big business with the rise of supermarkets.

As trade became more and more global, growers everywhere had to compete on price, especially orchardists. For example, canning costs had long been relatively high in New Zealand because of wage rates and small domestic runs. Growers supplying large processing companies in more populous countries could achieve such economies of scale that Hawke's Bay peaches and Central Otago apricots faced competition from South American and South African fruit. In those regions margins may have been thin, but they ran across huge volumes.

Before the 1950s, vegetable growing was largely confined to local market gardens selling through greengrocers and to home gardens. The bottling of fruit and making of jam were chores done annually by many housewives, but all this home processing changed as cities grew bigger and New Zealanders rapidly became urbanised in lifestyle.

The refrigerator with a freezer compartment earned its presence in every household during the two decades after the war and by 1955, Wattie's was quick-freezing peas in both Hastings and Gisborne.

The vegetable industry remained focused on the domestic market but counter-seasonal advantages allowed New Zealand growers to fill niches in other countries, especially the northern hemisphere, via fast and efficient transport systems, including containerised shipping. High-quality, long-keeping fresh vegetables such as onions, squash and potatoes became significant export earners, filling out-of-season gaps in local supplies in a number of countries.

Consumers in all developed countries abandoned the idea in the 1980s that seasons should be allowed to affect the supply of fresh produce, so New Zealand growers and packhouses began to export hydroponic lettuces, bagged salads, packaged greenhouse capsicums and tomatoes and other lightly processed ready-to-eat products, as well, of course, as frozen, dehydrated and canned vegetables.

Exports of fresh vegetables to dozens of countries by 2004 totalled nearly $220 million, with onions ($92 million) and squash ($53.5 million) the big winners, but with capsicums ($24 million), carrots ($10 million), tomatoes ($7 million and rising), and asparagus ($7 million) emerging with potential. New Zealand became the largest exporter of fresh onions to the European Community, and of squash (kabocha) to Japan. Processed vegetables earned $265 million mainly from beans ($40.3 million), potatoes ($66 million), peas ($39 million), sweetcorn ($50 million) and 'mixed vegetables' ($49 million).

Consumers in all developed countries abandoned the idea in the 1980s that seasons should be allowed to affect the supply of fresh produce, so New Zealand growers and packhouses began to export hydroponic lettuces, bagged salads, packaged greenhouse capsicums and tomatoes.

Cut flowers, mostly to Japan, pulled in $39 million, seeds, plants and bulbs $67 million, and if you wanted to count wine as a processed product of horticulture then it earned $302 million, and rising. Vegetables accessed markets throughout Australia and the Pacific islands, South East Asia, Korea and Japan.

The domestic market is worth many millions more to vegetable growers, though, and supply and distribution changed greatly towards the end of the twentieth century. Along with every other sector of primary industry, vegetable producers had to get bigger to survive. Fewer, larger growers were forced by the market to provide greater volumes of produce to fewer retail outlets, and directly, not through an intermediary. This caused the demise of the auction system in the early 1990s as supermarkets — reduced to just two chains in the late 1990s — consolidated their grip on the market.

The auction system, which a lot of the traditional growers valued, was too haphazard for the supermarkets. They wanted certainty and they wanted volumes. They wanted pallet loads rather than boxes and crates and they wanted supplies on 363 days a year; so the supply chain had to change and it changed dramatically. It didn't happen overnight. One of the auction floors was in action in Auckland for years after others closed, but that became an 'on-consignment' floor.

The effect on the growers was best demonstrated by the fact that in the early 1990s, the growers' federation had more than 5000 grower members. During the following decade the number of members shrank by half. A lot of the early suppliers were small and many of them part-timers. They would deliver their produce in a small van, hand it over to auctioneers and leave it with them to sell. When the auction system disappeared the small growers didn't have a customer any more; so they disappeared quite quickly in the 1990s.

What happened next was that committed growers became more committed, more commercial, bigger and the area in production went up. Yields were increasing at the same time because of the spread of artificial vegetative propagation involving tissue culture from the late 1980s. When growers planted seed, they would suffer a thirty or forty per cent crop loss, but with tissue culture-raised plants in the same area of ground they got a ninety per cent return. The result was crop yield boosts of around thirty per cent.

Global competition meant the pressure was always on growers. Markets for vegetable exports remained fragile in 2005 as buyers cast around the world for suppliers who would pitch their prices down to clinch a sale, prices which were always powerfully affected by seasonal weather conditions in growers' countries.

But New Zealanders remained dedicated to spreading the risk among a wide spectrum of produce with entrepreneurial moves into new varieties, with an explosion of growth in greenhouse salad vegetable production and by adapting produce established in other countries to local conditions. One Canterbury grower bred purple asparagus and export demand snapped up almost all his crop. Others moved into olives, producing tautologically tagged 'pure virgin' oil of top quality; and yet others experimented successfully with truffles.

Berryfruit also depend on local demand but blueberries, strawberries and boysenberries did have counter-seasonal export markets that varied from year to year, depending on the effect on prices of local production in those markets. Overall export returns have reached $14 million. Blueberries rode high in the new century in recent years on the back of health claims but, along with boysenberries and strawberries, they faced disease concerns. Blackcurrants went into broad acre production, particularly in Canterbury, and occupied a greater planted area (1300 hectares) than all other berryfruit combined, supplying processors and manufacturers of some of the most popular fruit drinks on the market.

As airline routes expanded and direct point-to-point flights became more common, the potential for exporting fresh, frozen and processed vegetables expanded. By the second decade of the twenty-first century, onions, squash and other fresh vegetables as well as processed peas, potatoes and sweetcorn and frozen vegetables earned more than $600 million.

## Chapter Twenty-Four

# A Dangerous Future is Crowding In

The first settled farmers in the Middle East and Asia sustained the first settled villagers within their region of occupation, a co-operation that lasted for a long time. Their agricultural practices may have been primitive compared with science-driven techniques of today, but life was rhythmically seasonal and enabled people to gather in stable groups, and turn their minds to other things, leading to an acceleration of intellectual and artistic life.

British cultural historian Peter Watson wrote of the Babylonians in his seminal *The History of Ideas*:

> *We call it civilization and we are apt to think of it as reflected in the physical remains of temples, castles and palaces that we see about us. But it was far more than that. It was a great experiment in living together, which sparked a whole new psychological experience, one that, even today, continues to excite many more of us than the alternatives. Cities have been the forcing house of ideas, of thought, of innovation, in almost all the ways that have pushed life forward.*

Cities also made settled, stable social groups more politically and economically manageable — and taxable, too, sometimes by local tyrants.

It is important to remember that the original cities Watson describes were tiny by today's standards, more intimately communal than the megacities of today and nestled within their rural regions. Attica — Athens and its rural surrounds — in the fifth century BC, the Age of Pericles, had around 200,000 people. The population of Athens at the centre was around 40,000, considerably less than Invercargill's today. Shakespeare's dynamic London of the late sixteenth century also had about 200,000 people, fewer than half as many as Wellington is home to today. (Rome and some Asian conurbations had populations estimated at around a million at most, but they were exceedingly rare.) So food and fibre production and consumption remained local in most of the world through most of recorded history with city and rural lives mainly integrated.

Megacities began to form along improving roads and railways, in the West anyway, in the nineteenth century which is one reason why people have had so little experience managing them. From then on transport systems expanded trade in farm products into first a national and then an international business.

Māori were always closely attached to the land and to the sea. While the Tāmaki isthmus may once have had a population that verged on urban, hapū and whānau were never far from their food sources. But agriculture ceased to be strictly local in Aotearoa from the time northern hemisphere plants and animals arrived, trees were replaced by pasture and migrants flooded in. Māori adopted the European plough even more readily once the horse became available. They quickly adapted the plants and animals that were brought into this country by the new arrivals.

By the last quarter of the nineteenth century, exports to Britain were fundamental to land use in New Zealand and a farming system was built around wool, lamb and dairy products. Rural life was pleasant enough among those who had capital and ran sheep for their wool in the Wairarapa, Marlborough and Canterbury, but work was mostly arduous, unremitting and in remote places often inaccessible by reliable roads for the smallholders around whom the dairy industry was formed.

As the wool market began to shrink from the 1960s, an increased variety of exports to an expanding number of nations followed Britain's entry to Europe in the 1980s. Gradually, after the Second World War, widened sealed roads made small towns and their regional farms easily accessible to cities. With more access to education and technology, the lives of farmers and their families became more urbanised. But the structure stayed basically intact — mostly family units applying as much technical expertise as they could command to produce food and fibre for export.

The moot point is how long this pattern could hold its shape and how adaptable the agricultural industry is in the face of change, because change it will. Farming and rural life around the world has been in flux in the new century, affected by population increases, especially in large cities, and rapidly changing cultures amid swirling international migrations; and also, by swelling flows of tourism.

Isolation kept New Zealand relatively safe from invasion from exotic animal and plant diseases. The biggest fear for most of the twentieth century was that foot and mouth disease would arrive and shut down exports of animal products. But recent incursions of Psa bacteria in kiwifruit stock, of *Mycoplasma bovis* into the cattle population, and increasingly frequent fruit fly arrivals demonstrate how porous borders must inevitably become as millions of visitors and millions of tonnes of freight pour into the country, and how costly it is to repel these invasions. France is fighting to confine a fungal grape disease among its grapes at a cost of billions of dollars. African swine fever, a highly contagious viral disease, is sweeping through Africa, Asia and, more recently, Europe, and millions of pigs are being culled in a bid to curb its spread.

But the most dramatic effect on New Zealand land use and production is likely to come, quite quickly now, from the impact of climate change and the consequences of striking variations in the national diet.

Nutritional studies have been notoriously unreliable, often contradictory, burdened by spurious detail and a failure to acknowledge national and individual differences. As examples, fluctuating results come from a myriad of 'studies' on the damage that may be done or not by alcohol in even small amounts. They are accompanied with varying claims surrounding the danger of eating red and processed meat.

However, some broad certainties have arisen, and many young people seem to accept them. A daily drink of beer or wine without regular breaks and associated with occasional binge drinking is undoubtedly detrimental to the good health of most people. How likely is it that this will disrupt the areas planted in barley and grapes?

Even more secure is the understanding that regular, often daily, consumption of red meat, bacon and other preserved meats will affect production of these foods as the evidence firms up that it is seriously detrimental to the health and longevity of most of us.

Disciplined eating and drinking along with exercise are becoming part of the daily regime of young people and farmers will ignore this trend at their peril. There is already a falling off of lamb consumption and some farmers are adjusting production towards the top end of the market to catch the occasional foodie in search of red meat.

Match all this with the knowledge that many soils now covered with grass will be more productive under well-managed horticultural regimes than they are under pastoral management. If you think dietary trends take time, it is worth recalling that changes of food preference in the decades after the Second World War were among the most rapid and complete social and cultural changes in Aotearoa last century.

More direct and urgent, though, is growing apprehension about climate change. Denial that the activities of modern people are accelerating any change represents a failure of the imagination of anyone who stands above motorways in central Auckland, or any modern expanding city.

Shortages of water are already impeding traditional farm production in a number of countries, including Australia. The *New York Times* reported early in 2019: 'In every region, farmers and scientists are trying to adapt an array of crops to warmer temperatures, invasive pests, erratic weather and earlier growing seasons.' As farm sizes have built up with larger and larger areas in single crops, swift adaptation to these weather changes has become more difficult. I recall driving in Iowa a few years ago and for kilometre after kilometre maturing corn lined each side of the road. Soon, I knew, contractors would move in for the harvest.

More than ninety per cent of all soybean and corn meal production in the United States comes from this sort of factory farm. Most of it goes to feedlots for the factory farming of beef and dairy cattle, pigs and chickens. This agglomeration of one-time family farms into corporations has economically and socially disrupted American rural life. An even greater disruption is likely as climate change gains momentum. The same pattern of huge areas in single crops, mostly soybeans and corn, has changed rural lives in countries like Argentina where failures in genetically modified crops are also causing problems.

Huge areas under factory production are involved in large agricultural countries. Changing land use and crops in many of them will be like turning a liner around in a river if climate change accelerates and the public clamour for action escalates.

Not much more than half of New Zealand is farmland and that half is divided into about 55,000 holdings. Three-quarters of that land is in pasture. As the second decade of the twenty-first century closed these farms carried about twenty-seven million sheep, about 3.5 million beef cattle, 6.5 million dairy cows and more than a million deer. Many of these animals, some a lot more than others, contribute a third of New Zealand's greenhouse gas emissions.

The main fruit crops include grapes (producing 2.8 million hectolitres of wine), kiwifruit, apples and avocados, all attracting growing export earnings. Large areas, mostly in Canterbury, carry wheat, barley, oats, maize and pasture seeds, much of it to feed animals as well as people.

None of these numbers are static and are likely to become more volatile in coming years, even though they may be more stable than in many other countries. A crumpled country with a number of microclimates conditioned by rainfall, Aotearoa lacks long level stretches of land which tends to make the family farm on a smaller scale a better fit even today; although farm size has been scaled up among corporate dairy farms on the east coast of the South Island. The planting and harvesting of exotic forests for export by foreign owners is certainly factory farming of a sort. Could these trees become the base of a biofuel industry or expanded to balance carbon quotas?

Around the world, change to both urban and rural living has been exponentially rapid since the 1980s and is continuing, which makes peering into the future difficult, but the future is crowding in on farmers and on rural towns and it will need an alert and flexible industry to cope.

## References


*Hints to Intending Sheep-Farmers in New Zealand,* by Frederick Weld, 1851.

*The Story of New Zealand,* by A. S. Thomson, 1859.

*A First Year in the Canterbury Settlement,* by Samuel Butler, 1863.

*Station Life in New Zealand* by Lady Barker, 1870.

*On the Maori Races of New Zealand,* a collection of papers read to the Hawkes Bay Philosophical Institute between 1878 and 1882 by William Colenso.

*The Long White Cloud,* William Pember Reeves, 1898.

*Tutira: The Story of a New Zealand Sheep Station* by H. Guthrie-Smith, 1921.

*Maori Agriculture* (Dominion Museum, Bulletin No. 9, 1925), by Elsdon Best.

*The Early Canterbury Runs,* by L. G. D. Acland, 1930.

*Follow the Call* by Frank S. Anthony, 1936.

*High Country – the Evolution of a New Zealand Sheep Station* by R. M. Burdon, 1938.

*Wheat in New Zealand,* by F. W. Hilgendorf, 1938.

*Brave Days: Pioneer Women of New Zealand,* a collection of pioneering stories written by women for the NZ Centennial, 1940.

*The Farmer in New Zealand* by G. T. Alley and D. O. W. Hall, 1941.

*Early Runholding in Otago* and *The Pioneers Explore Otago,* by Herries Beattie, 1947.

*John Grigg of Longbeach,* by Percy G. Stevens, 1952

*Grasslands of New Zealand*, by Bruce Levy, 1955.

*Wool Away,* by Godfrey Bowen, 1955.

*Principles of Animal Production* and *Grass to Milk,* by C. P. McMeekan, 1960.

*The Age of the Great Sheep Runs* P. R. Stephens, 1965.

*Farm and Station Verse 1850-1950* collected by A. E. Woodhouse, 1967.

*Challenge and Response: A History of the Gisborne East Coast Region* by W. H. Oliver, 1971.

*Arable Farm Crops of New Zealand,* by J. H. Claridge, 1972 (a revision of J. W. Hadfield's book of the same title in 1952, twenty years previously).

*Golden Jubilee: NZ Meat Producers Board, First Fifty Years,* edited by Dai Hayward, 1972.

*A Command of Co-Operatives*, by A. H. Ward, 1975.


It is difficult to overestimate the historical resource of the *New Zealand Journal of Agriculture* and *The New Zealand Year Book,* both published by governments through most of the 20th century.

# Index

## THE AUTHOR

GORDON MCLAUCHLAN is a highly regarded author, journalist and social commentator, writing for newspapers and magazines both locally and internationally, and is the author or contributing editor of more than twenty books, including *A Short History of New Zealand* and *A Short History of the New Zealand Wars.* His specific areas of interest are the historical, social and political heritage of New Zealanders, literature, agriculture, tourism and media.

In the course of his journalistic career he has worked as a senior reporter, sportswriter and sub-editor on provincial and regional newspapers, as well as for the New Zealand Broadcasting Service and the New Zealand Press Association. Since the mid-1970s he has been a freelance writer, broadcaster and public affairs consultant, including nine years as an investigative reporter for the *National Business Review*, and as books editor of the *New Zealand Herald* for which he also wrote a weekly op-ed column for thirty years.

A generalist with a wide range of interests, Gordon has a substantial private reference library, which was well utilised when he edited and wrote much of Bateman's *New Zealand Encyclopedia*, and compiled the 2000 New Zealand questions for *Trivial Pursuit* as well as the questions for the first two seasons of the television programme, *Sale of the Century.*

He has had a long association with the New Zealand Society of Authors (PEN NZ Inc.), including a two-year stint as president (1995–96), and also as the New Zealand delegate at International PEN congresses in Edinburgh, Helsinki, Warsaw and Moscow. In addition, Gordon was the founding president of the Travel Communicators of New Zealand (Travcom); the founding chair of the Michael King Writers' Studio Trust; has been a trustee of the Frank Sargeson Trust for twenty years; and took over as editor of the New Zealand Automobile Association's *Directions* magazine from 1998–99.

Gordon served as a Wattie Book of the Year judge in 1978 and a convener in 1979; a Montana National Book Awards judge in 1998 and convener in 1999; and was a director for seven years (three of them as chairman) of Copyright Licensing Limited (now Copyright Licensing New Zealand) which controls licences and revenue for copyright holders and publishers. In 2019 he was appointed an Officer of the NZ Order of Merit for his services to historical research.